OPTIMIZATION

Linear Programming

OPTIMIZATION
Linear Programming

B.N. Mishra
B.K. Mishra

Ane Books Pvt. Ltd.
New Delhi ♦ Chennai

Optimization: Linear Programming

B.N. Mishra and B.K. Mishra

First Edition : 2012

Reprint : 2018, 2019, 2020, 2023

Published by

Ane Books Pvt. Ltd.

4821, Parwana Bhawan, 1st Floor, 24 Ansari Road, Darya Ganj,
New Delhi - 110 002, Tel.: +91(011) 23276843-44, Fax: +91(011) 23276863
e-mail: kapoor@anebooks.com, Website: www.anebooks.com

Branch Office:

- Avantika Niwas, 1st Floor, 19 Doraiswamy Road, T. Nagar,
Chennai - 600 017, Tel.: +91(044) 28141554, 28141209
e-mail: anebookschennai@gmail.com, rathinam@anebooks.com

ISBN : 978-81-8052-125-6

Printed at : Repro Knowledgecast Limited, Thane

Preface

The book entitle “Linear Programming” has been written to serve suitable text book for the students of undergraduate Mathematics/Statistics/Management/ Economics class whose courses of studies include some or all the topics treated here.

The subject “Linear Programming” is very modern in nature. It forms a part of “Operation Research” which was first coined in 1939 by Mc-Closky and Trefthen in a small town. Bowdsey of United Kingdom. This new science comes into existence in military context. During World War II, military management sought the co-operation of scientists of various disciplines and organized them into teams to assist in solving strategic and tactical problems. This led to remarkable progress and success in execution of military operation and project. Following the end of the war, the success of it arrested the attention of industrial managers who were seeking solutions of their complex economic problems. Since then the subject has been used to achieve the optimization of available resources. This made a direct impact on the studies of industrial and economic problems. The scientific approach to decision making problems simplified the situation to a great extent. Really speaking, this subject is the attack of modern science on problems of likelihood that arisen in management and control of resources in their natural environment.

This subject has special technique to attack the problems. But we have arrived to make it simple and systematic with the help of numbers of illustrative examples. For this we have consulted a number of authors freely for which we are grateful to them.

We shall thankfully receive constructive suggestions for its improvements from all those interested in studying this subject.

B.N. Mishra
B.K. Mishra

Contents

1

Linear Programming

1.1. INTRODUCTION

Linear programming is a subject of recent origin. But it has already demonstrated its value as an aid to decision making in business, industry and government. Determination of facility or machine scheduling, distribution of commodities, determination of optimum product mix and allocation of labour and other resources are but a few examples of the type of problems which can be solved by linear programming. Many business and economic are concerned with a problem of planning activity. In such case there are limited resources at the disposal and the problem is to make such a use of these resources so as to yield the maximum production or to minimise the cost of product, or to give the maximum profit etc. Such problems are referred to as problems of constrained optimization (maximization or minimization). Briefly *linear programming is a technique for determining an optimum programme of interdependent activities in view of available resources*. The term "*linear*" stands for indicating that all relationships involved in a particular problem which can be solved by this method are linear. The term "*Programming*" is just another word for planning and refers to the process of determining a particular plan of action from amongst several alternatives.

It is one of the most important optimization techniques developed in the field of "*Operation Research*". The reason why linear programming asserts so much attention today is that methods of solution to the general linear programming problem involving n variables have been developed. The best known and most widely used of these methods is called the "*Simplex method*" which was developed by George Dantzig in 1947.

1.2. MATHEMATICAL FORMULATION OF LINEAR PROGRAMMING PROBLEMS (L.P.P.)

A liner programming problem includes a set of simultaneous linear equations (or inequalities) which represent the conditions of the problems and a linear function which expresses the objective function of the problem. The linear function which is to be optimized is called the *objective function* and the conditions of the problems expressed as simultaneous linear equations (or inequalities) are referred as *constraints* (or restriction).

These two terms can be explained by the following example

Maximize $Z = 6x_1 + 11x_2$ ... ... (i)

Subject to $2x_1 + x_2 \le 104$ ... ... (ii)

$x_1 + 2x_2 \le 76$ ... ... (iii)

and $x_1 \ge 0$, $x_2 \ge 0$ ... ... (iv)

(i) is objective function, (ii) and (iii) are constraints. Also, (iv) is regarded as condition of non-negativity.

Now, we proceed to translate a linear programming problem into the language of algebra (that is algebraic equations, inequalities and expressions). The following problems of linear programming will help us in understanding the conversion work.

Example 1—A manufacturer of line of patent medicines is preparing a production plan on medicines A and B. There are sufficient in gradients available to make 20,000 bottles of A and 40,000 bottles of B but there are only 45,000 bottles into which either of the medicines can be put. Furthermore, it takes 3 hours to prepare enough material to fill 1,000 bottles to A, it takes one hour to prepare enough material to fill 1,000 bottles of B and there are 66 hours available for this operation. The profit is Rs. 8/- per bottle for A and Rs. 7/- per bottle of B.

(a) Formulate this problem as a liner programming problem (L.P.P.).

(b) How should the manufacturer schedule production in order to maximize his profit? [Meerut O.R. '73]

Solution: (a) Let the manufacturer produce x_1 and x_2 bottles of medicines A and B respectively. Let the total profit (in Rs.) be z.

Then $$z = 8x_1 + 7x_2 \quad ...(1)$$

The time required to prepare x_1 bottles of medicine A is $\frac{3x_1}{1000}$ hours and the time required to prepare x_2 bottles of medicine B is $\frac{1x_2}{1000}$ hours.

∴ total time required to prepare x_1 bottles of medicine A and x_2 bottles of medicine B is $\left(\frac{3x_1}{1000}+\frac{x_2}{1000}\right)$ hours

Since total time available for this operation is 66 hours, therefore

$$\frac{3x_1}{1000}+\frac{x_2}{1000} \le 66$$

or, $3x_1 + x_2 \le 66{,}000$

Since there are only 45,000 bottles into which the medicines can be put therefore $x_1 + x_2 \le 45{,}000$

Hence, the problem of linear programming can be put in the following mathematical form:

Maximize $z = 8x_1 + 7x_2$				... (i)
Subject to	$3x_1 + x_2$	$\le$	66,000	... (ii)
	$x_1 + x_2$	$\le$	45,000	... (iii)
	x_1	$\le$	20,000	... (iv)
	x_2	$\le$	40,000	... (v)
and	x_1	$\ge 0,\ x_2 \ge 0$		... (vi)

The above is known as mathematical formulation of the given linear programming problem (i) is objective function, (ii) to (vi) are constraints (restriction). This problem is called a linear programming problem (L.P.P.), since a linear (first degree) function of two variables x_1 and x_2 defined by $z = 8x_1 + 7x_2$ is to be maximized under the linear constraints (i) to (v).

(b) For solution of the above problem see Example 4.

SOME IMPORTANT DEFINITIONS

(1) **Feasible Solution**—Any set $x = \{x_1, x_2, \ldots x_{n+m}\}$ of variables is called a feasible solution (or programme) of the Linear Programming problem, *if it satisfies the constraints and non-negativity restrictions.*

(2) **Optimum Solution**—A feasible solution of a L.P.P. is said to be optimum (or optimal) solution *if it also optimizes the objective function Z of the problem.*

(3) **Unbounded Solution**—If the value of the objective function Z can be increased or decreased indefinitely, then such solutions are called Unbounded Solutions.

Methods for the Solution of a L.P.P.

In general, the following two methods are adopted for the solution of a Linear Programming Problem.

(i) **Geometrical or Graphical Methods**—If the objective function Z is a function of two variables only, then the L.P.P. can be solved effectively by the graphical method.
If Z is a function of three variables, then also it can be solved by this method. But in this case the graphical solution becomes complicated enough.

(ii) **Simplex Method**—This is the most powerful tool of the linear programming as any problem can be solved by this method. This method is an algebraic procedure which progressively approaches the optimum solution.

Graphical Method for the Solution of L.P.P.

For the graphical solution the following steps are used:

(i) **Step one**—Consider the constraints as equalities.

(ii) **Step two**—Draw the lines corresponding to each equation and non-negative restrictions.

(iii) **Step three**—Obtain the permissible region for the values of the variables, which is the region bounded by the lines drawn in step two.

(iv) **Step four**—Find a point in the permissible (or feasible) region which gives the optimum value of the objective function Z. For this the use of the following proposition can be made. *If S be a convex subset of the plane bounded by the lines in the planes, then a linear function $Z = c_1x_1 + c_2x_2$, $\forall (x_1, x_2) \in S$ where c_1, c_2 are scalars attains its extreme value (maximum or minimum) at the vertices.*

Limitations of linear programming—The following will be the assumptions of linear programming problem that limits its applicability.

1. **Proportionality**—A primary requirement of linear programming problem is that the objective function and every constraint must be linear. Roughly speaking, it simply means that 1 kg of a product costs Rs. 5, 10 kg will cost Rs. 50.
2. **Additivity**—Suppose we use T_1 hours on machine A to make product P_1 and T_2 hours to make product P_2, then the time on machine A devoted to products A_1 and P_2 both is $T_1 + T_2$ provided the time required to convert from product P_1 and P_2 is negligible.
3. **Multiplicativity**—The total profit from selling a given number of units is the unit profit times the number of units sold.
4. **Divisibility**—It means that the fractional levels of variable must be permissible besides integral values.
5. **Deterministic**—All the coefficients in the linear programming models are assumed to be known exactly.

A General Linear Programming Problem:

[R.U.B.Sc. 85, 86] (V.B.U. Part III Hons 97 (Old spl.))

A L.P.P. for n variables $x_1, x_2 \ldots x_n$ can be stated as follows. Find $x_1 \ldots x_2 \ldots x_n$ which Optimize the linear function

$$Z = c_1x_1 + c_2x_2 + \ldots + c_nx_n \qquad \ldots(1)$$

Subject to the constraints

$$\left.\begin{array}{l} a_{11}x_1 + a_{12}x_2 + \ldots + a_{1n}x_n \ (\leq = \geq)\ b_1 \\ a_{21}x_1 + a_{22}x_2 + \ldots + a_{2n}x_n \ (\leq = \geq)\ b_2 \\ \ldots \quad \ldots \quad \ldots \quad \ldots \quad \ldots \quad \ldots \quad \ldots \quad \ldots \quad \ldots \\ a_{m1}x_1 + a_{m2}x_2 + \ldots + a_{mn}x_n \ (\leq = \geq)\ b_m \end{array}\right\} \qquad \ldots(2)$$

and non-negative restrictions $x_1, x_2, \ldots x_n \geq 0$ ...(3)

where all a'', b'' and c'' are constants. In the constraints (2) there may be any of the three signs $\leq, =, \geq$.

The function Z given by (1) is called the objective function and the conditions given by (2) are called *constraints* of the linear programming problem.

The above linear programming problem can be put in the matrix form as below:

Optimize $Z = cx$

Subject to $Ax\ (\leq, =, \geq)\ b$

and $x \geq 0$

$$A = \begin{bmatrix} a_{11} & a_{12} \cdots\cdots & a_{1n} \\ a_{21} & a_{22} \cdots\cdots & a_{2n} \\ \cdots\cdots & \cdots\cdots\cdots & \\ a_{m1} & a_{m2} & a_{mn} \end{bmatrix}$$

$c = (c_1, c_2, \ldots c_n)$ is a row vector known as price vector

$$x = \begin{bmatrix} x_1 \\ x_2 \\ \cdots \\ x_m \end{bmatrix} = [x_1, x_2, \ldots x_n]^T$$

is column vector of variables

$$b = \begin{bmatrix} b_1 \\ b_2 \\ \cdots \\ b_m \end{bmatrix} = [b_1, b_2, \ldots b_m]^T$$

is column vector called the requirement vector.

2. Solve Max. to $Z = 5x_1 + 7x_2$

Subject to $3x_1 + 2x_2 \leq 12$

$2x_1 + 3x_2 \leq 13$

$x_1, x_2 \geq 0$ by graphical method

Solution: Taking equality signs the constraints become.

$$3x_1 + 2x_2 = 12 \qquad ...(1)$$
$$2x_1 + 3x_2 = 13 \qquad ...(2)$$
$$x_1 = 0, x_2 = 0 \qquad ...(3)$$

which represent straight lines being first degree.

$x_1 \geq 0$ and $x_2 \geq 0$ suggest that the feasible solution is possible only in the first quadrant. Draw the lines given by the equations (1), (2) and (3). Lines drawn are shown in figure. By solving

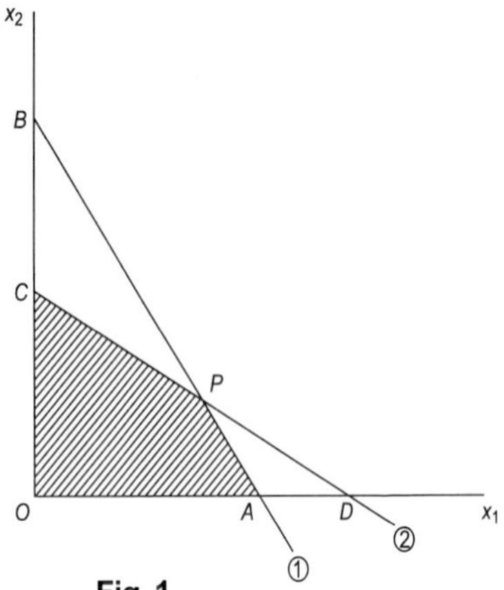

Fig. 1

$$3x_1 + 2x_2 = 12 \qquad ...(1)$$
and $$x_2 = 0$$

we get their point of intersection $A(4, 0)$

Solve $$2x_1 + 3x_2 = 13 \qquad ...(2)$$
and $$x_1 = 0$$

to get the co-ordinates of their point of intersection C.

C is $\left(0, \dfrac{13}{3}\right)$. P is obtained by solving

$3x_1 + 2x_2 = 12$ (1) and $2x_1 + 3x_2 = 13$ (2)

P is (2, 3)

The region bounded by these four lines (shown as shaded) is the permissible (or feasible) region of solution. Values of Z at the vertices will give either maximum or minimum value.

Value of Z at $A(4, 0)$ is $5 \times 4 + 7 \times 0 = 20$

Value of Z at $P(2, 3)$ is $5 \times 2 + 7 \times 3 = 31$

Value of Z at $C\left(0, \dfrac{13}{3}\right)$ is $5 \times 0 + 7 \times \dfrac{13}{3} = \dfrac{91}{3}$

Value of Z at $O(0, 0)$ is 0

Clearly, the maximum value of Z is 31

at $x_1 = 2, x_2 = 3$

3. Solve by graphical method

Min.$Z = 1.5x_1 + 2.5x_2$

Subject to $x_1 + 3x_2 \geq 3$

$x_1 + x_2 \geq 2$

$x_1 \geq 0, x_2 \geq 0$

Solution: Using equality signs in the constraints, we plot the lines $x_1 + 3x_2 = 3$; $x_1 + x_2 = 2$ and then shade the feasible region as shown in the figure. The minimum value of Z is Z min. (= 3.5). This minimum is attained at the point of intersection A of the lines $x_1 + 3x_2 = 3$ and $x_1 + x_2 = 2$. This is unique point to give the minimum value of Z.

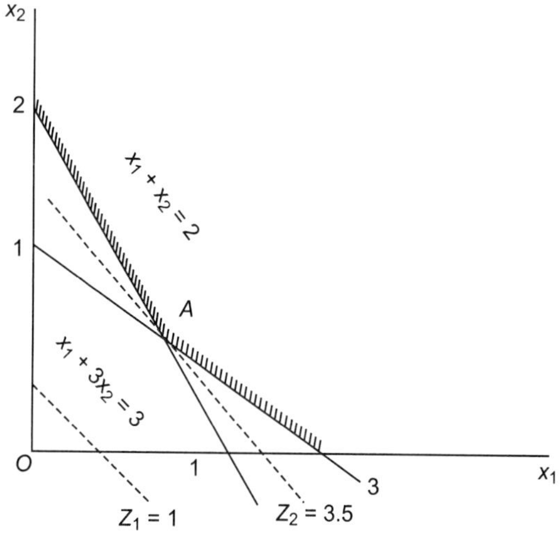

Fig. 2

Solving these two equations we get $x_1 = \frac{3}{2}$, $x_2 = \frac{1}{2}$

$$Z = f(x_1, x_2) = (1.5)x_1 + (2.5)x_2$$

$$\text{Min } Z = f\left(\frac{3}{2}, \frac{1}{2}\right) = (1.5)\,\frac{3}{2} + (2.5)\,\frac{1}{2}$$

$$= 3.5 \text{ by graphical method}$$

4. Solve the following programming problem by graphical method

$$\text{Max. } Z = 8x_1 + 7x_2$$

$$\begin{aligned} \text{Subject to} \quad & 3x_1 + x_2 \le 66{,}000 \\ & x_1 + x_2 \le 45{,}000 \\ & x_1 \le 20{,}000 \\ & x_2 \le 40{,}000 \\ & x_1 \ge 0, x_2 \ge 0 \end{aligned}$$

[Meerut O.R. 73]

Solution: Firstly, we convert the inequalities into equation as follows

$$3x_1 + x_2 = 66{,}000 \qquad ...(1)$$

$$x_1 + x_2 = 45{,}000 \qquad ...(2)$$

$$x_1 = 20{,}000 \qquad ...(3)$$

$$x_2 = 40{,}000 \qquad ...(4)$$

Then the lines are drawn and shown in the adjoining figure.

$$Z = 8x_1 + 7x_2 = 0$$
$$8x_1 = -7x_2$$

or, $$\frac{x_1}{-7} = \frac{x_2}{8} = t$$
$$x_1 = -7t$$
$$x_2 = 8t$$

The shaded region shown in the fig. 3 is the permissible region for the value of the variables x_1 and x_2.

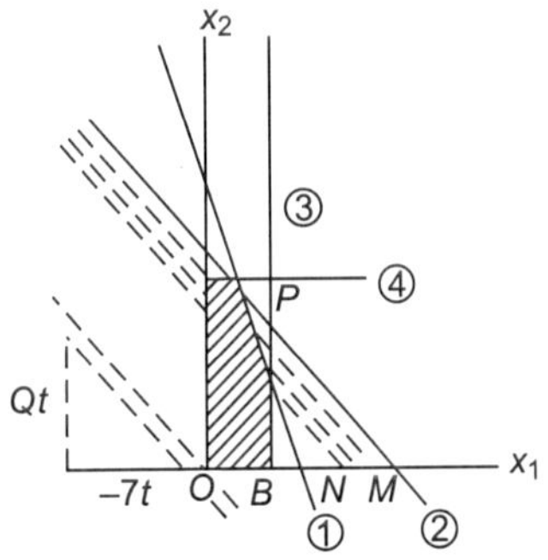

Fig. 3

To find x_1, x_2, for which Z is maximum, we draw the line (dotted line through O) $Z = 8x_1 + 7x_2 = 0$ and continue drawing parallel lines till we reach the point of the permissible region which is farthest away from the origin.

As shown in the fig. this point is at point $P(10{,}500; 34{,}500)$ which is the point of intersection of lines.

$$3x_1 + x_2 = 66{,}000$$

and $$x_1 + x_2 = 45{,}000$$

Therefore, Z is maximum for $x_1 = 10{,}500$ and $x_2 = 34{,}500$ i.e. $Z = 8(10{,}500) + 7(34{,}500) = 3{,}25{,}500$ *Ans.*

5. Solve the following L.P.P. by graphical method

Max. $z = 6x_1 + 11x_2$

Subject to $2x_1 + x_2 \le 104$

$x_1 + 2x_2 \le 76$

$x_1 \ge 0, x_2 \ge 0$

[R.U.B. Sc. 85; V.B.U. Part III Hons '99]

Solution: First change the constraints as equations

$$2x_1 + x_2 = 104 \quad ...(1)$$
$$x_1 + 2x_2 = 76 \quad ...(2)$$
$$x_1 = 0,\ x_2 = 0$$

$x_1 \ge 0$, and $x_2 \ge 0$ suggest that the feasible region is the first quadrant.

Draw the lines given by the equation. Lines drawn are shown in the figure. The feasible region is shown as shaded. Find the co-ordinates of M, P and A

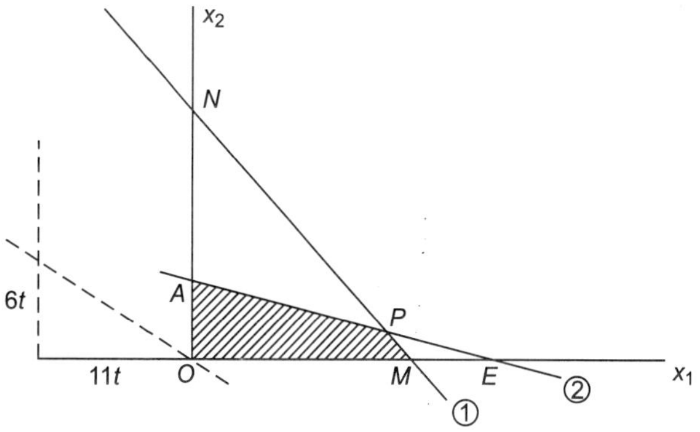

Fig. 4

$$= 6x_1 + 11x_2 = 0$$

$$\frac{x_1}{11} = \frac{x_2}{6} = t$$

Solving $\quad 2x_1 + x_2 = 104 \quad$ ...(1)

and $\quad x_1 + 2x_2 = 76 \quad$ ...(2)

we get $\quad x_1 = 44, \quad x_2 = 16$

P is $\quad (44, 16)$

Similarly, we can obtain M and A. By putting the co-ordinates of A, P and M in $Z = 6x_1 + 11x_2$ we see the maximum value is at P.

$$Z \text{ max.} = 6 \times 44 + 11 \times 16 = 440.$$

6. A manufacturer produce two types of models M_1 and M_2. Each M_1 model requires 4 hours of grinding and 2 hours of polishing; whereas each M_2 model requires 2 hours of grinding and 5 hours of polishing. The manufacturer has 2 grinders and 3 polishers. Each grinder works for 40 hours a week and each polisher works for 60 hours a week. Profit on an M_1 model is Rs. 3.00 and on an M_2 model is Rs. 4.00. Whatever is produced in a week is sold in the market. How should the manufacturer allocate his production capacity to the two types of models so that he may make the maximum profit in a week.

Solution: **Formulation of the mathematical problem.**

Let x_1 be the number of M_1 models and x_2 the number of M_2 models that the manufacturer decides to produce per week.

Then this weekly profit is given by

$$Z = 3x_1 + 4x_2$$

This is on the assumption that he has enough grinder and polisher capacity to produce these number of the models.

Now in order to produce these number of models, the total number of grinding hours needed per week is given by $4x_1 + 2x_2$, and the total number of polishing hours needed per week is given by $2x_1 + 5x_2$. Since he does not have more than 80 hours of grinding nor does he has more than 180 hours of polishing, therefore, we must have

$$4x_1 + 2x_2 \le 80$$

and $$2x_1 + 5x_2 \le 180$$

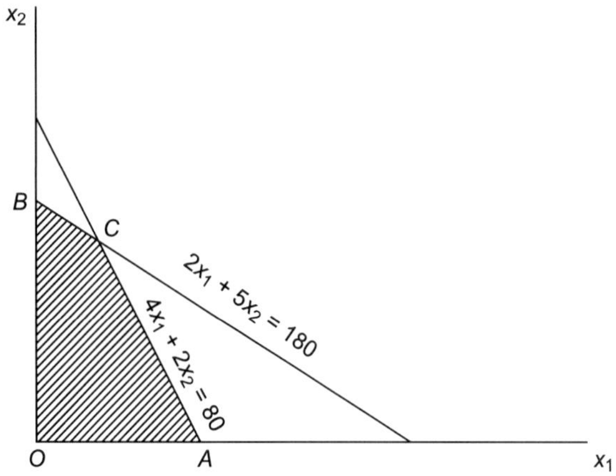

Fig. 5

It is clear that any point which satisfies the conditions

$$\left.\begin{aligned} x_1 &\ge 0 \\ \text{and} \quad x_2 &\ge 0 \end{aligned}\right\}$$

lies in the first quadrant and conversely for any point (x_1, x_2) in the first quadrant $x_1 = 0$ and $x_2 \ge 0$. Thus the number pair (x_1, x_2) is restricted to the points of the first quadrant only.

Similarly, since

$$4x_1 + 2x_2 \le 80$$

and $$2x_1 + 5x_2 \le 180$$

the desired point (x_1, x_2) must be somewhere in the convex shaded region *OACB* as shown in the fig. The convex region *OACB* is bounded by the lines,

$$\left.\begin{aligned} x_1 &= 0 \\ x_2 &= 0 \\ 4x_1 + 2x_2 &= 80 \\ \text{and} \quad 2x_1 + 5x_2 &= 180 \end{aligned}\right\}$$

This region is called the *solution space for the given problem*. The four vertices of the convex region are

$$O = (0, 0)$$
$$A = (20, 0)$$
$$B = (0, 36)$$
$$C = (2.5, 35)$$

Now,

Let $$Z = f(x_1, x_2) = 3x_1 + 4x_2$$

Then $$Z(0) = f(0, 0) = 3.0 + 4.0 = 0$$

Where $Z(0)$ stands for value of Z at the point O which can be obtained by putting the co-ordinates of O in

$$Z = f(x_1, x_2) = 3x_1 + 4x_2$$
$$Z(A) = f(20, 0) = 3 \times 20 + 4 \times 0 = 60$$
$$Z(B) = f(0, 36) = 3 \times 0 + 4 \times 36 = 144$$
$$Z(C) = f(2.5, 35) = 3 \times 2.5 + 4 \times 35 = 147.5$$

The maximum value of Z (viz. 147.5) occurs at

$C = (2.5, 35)$ and the minimum occurs at $O(0, 0)$.

7. A resourceful house decorator manufactures two types of lamp X and Y. Both lamps go through two technicians, first a cutter, second a finisher.

Lamp X requires 2 hours of the cutter's time and 1 hour of the finisher's time. Lamp Y requires 1 hour of cutter's and 2 hours of finisher's time. The cutter has 104 hours and finisher 76 hours of time available each month. Profit on one lamp X is Rs. 6 and on Y lamp is Rs. 11. Assuming that he can sell all that he produces, how many of each type lamps should be manufactured to obtain the best return.

Solution: Let the decorator manufacture x_1 and x_2 lamps of type X and Y respectively.

Therefore, total profit (in Rs.) $Z = 6x_1 + 11x_2$

Total time to the cutter used in preparing x_1 lamps of types X and x_2 of type Y is $2x_1 + x_2$.

Since cutter has 104 hours only for each month therefore

$$2x_1 + x_2 \leq 104$$

Similarly, the total time of the finisher used in preparing x_1 lamps of type X and x_2 of type Y is $x_1 + 2x_2 \leq 76$.

Hence is the decorator problem is to find x_1, x_2 which maximize,

$$Z = 6x_1 + 11x_2$$

Subject to the constraints

$$2x_1 + x_2 \leq 104$$
$$x_1 + 2x_2 \leq 76$$

and $$x_1 \geq 0, x_2 \geq 0.$$

For the solution see example 5

8. Product mix problem

A pineapple firm produces tow products-canned pineapple and canned juice. The specific amounts of material, labour and equipment required to produce each product and the availability of each of these resources are shown in the table given below:

	Canned juice	*Canned Pineapple*	*Available Resources*
Labour (man-hours)	3	2.0	12.0
Equipment (machine hours)	1	2.3	6.9
Material (unit)	1	1.4	4.9

Assuming one unit each of canned juice and canned pineapple has profit margins of Rs. 2 and Rs. 1 respectively, the problem is to determine the product mix that will maximize the profit.

Solution:

Mathematical Formulation:

Let the amounts of canned juice and pineapple to be produced be x_1 and x_2 respectively. Then for each unit of canned juice and pineapple the producer wishes to produce canned juice and pineapple in such amounts and combinations that the profit will be maximized i.e. his problem is to maximize.

$$Z = 2x_1 + x_2$$

Now since three man-hours are required to produce one unit of canned juice and two man-hours to produce one unit of canned pineapple, we need $3x_1 + 2x_2$ man-hours to produce x_1 units of canned juice and x_2 units of canned pineapple. These man-hours are limited to 12; thus we have

$$3x_1 + 2x_2 \le 12$$

This is the first inequation. Similarly, we can obtain the second and third inequations. Thus the three inequations are

$$3x_1 + 2x_2 \le 12.0$$
$$x_1 + 2.3x_2 \le 6.9$$
$$x_1 + 1.4x_2 \le 4.9$$

Also it is not possible for the producer to produce the negative number of unit of canned juice and pineapple, therefore, we must have

$$x_1 \ge 0 \text{ and } x_2 \ge 0$$

Hence this product mix problem can be solved with the following mathematical form i.e.

$$3x_1 + 2x_2 \le 12.0$$
$$x_1 + 2.3x_2 \le 6.9$$
$$x_1 + 1.4x_2 \le 4.9$$
$$x_1, x_2 \ge 0$$

And for which the expression (objective function)

$$Z = 2x_1 + x_2$$

where x_1 and x_2 be two real numbers, may be maximum.

Graphical Solution:

Let us consider a set of rectangular cartesian axis Ox_1x_2 in the plane. We see that any point (x_1, x_2) which satisfies the conditions

$$x_1 \geq 0 \;\&\; x_2 \geq 0$$

lies in the first quadrant only. Conversely for any point (x_1, x_2) in the first quadrant $x_1 \geq 0, x_2 \geq 0$. Now the number pair (x_1, x_2) is thus restricted to the points of the first quadrant only.

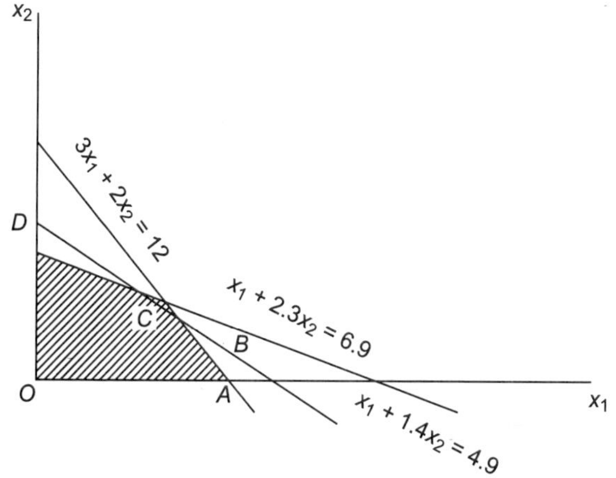

Fig. 6

Now since,

$$3x_1 + 2x_2 \leq 12.0, \qquad x_1 + 2.3x_2 \leq 6.9$$

$$x_1 + 1.4x_2 \leq 4.9$$

the desired point must be some where in the shaded region *OABCD* shown shaded in the Fig. 6

Since our search is for such a number pair (x_1, x_2) which give a maximum value of z the desired point will be at any one of the four vertices of the shaded region. Solving the equations, we get

O (0, 0); A = (4, 0); B = (3.2, 1.2), C = (1.7, 2.2) & D = (0, 3)

Values of the objective functions at these five vertices are: $Z(0) = 0$; $Z(A) = 8$; $Z(B) = 7.6$; $Z(C) = 5.6$ and $Z(D) = 3$. Obviously the maximum values of Z is 8 which occurs at A = (4.0). Hence the solution to the mix problem, $x_1 = 4$, $x_2 = 0$, max. $Z = 8$.

9. Diet Problem

Consider two different types of food stuffs, say, F_1 and F_2. Assume that these foodstuffs contain vitamins v_1, v_2 and v_3 respectively minimum daily requirements of these vitamins are 1 mg of v_1, 50 mg of v_2 and 10 mg of v_3. Suppose that the foodstuff F_1 contains 1 mg of v_1, 100 mg of v_2 and 10 mg of v_3 where as foodstuff

F_2 contains 1 mg of v_1, 10 mg of v_2 and 100 mg of v_3. Cost of one unit of foodstuff F_1 is Re. 1 and that of F_2 is Rs. 1.5.

Find the minimum cost diet that would supply the body at least the minimum requirements of each vitamin.

Solution: *Mathematical Formulations*

The following table will help us in solving the problem

	v_1	v_2	v_3	*Cost in Rs.*	*unit of food*
F_1	1	100	10	1	x_1
F_2	1	10	100	1.5	x_2
vitamin required	1	50	10		

Let us introduce two variable x_1 and x_2 to stand for units of foodstuffs F_1 and F_2 respectively.

Therefore, the total cost (in rupees) or diet consisting of x_1 units of F_1 and x_2 units of F_2 is given by

$$Z = x_1 + 1.5x_2$$

Now the foodstuffs F_1 and F_2 contain

total number of vitamins v_1 (*in* mg) equal to $x_1 + x_2$

total number of vitamins v_2 (*in* mg) equal to $100x_1 + 10x_2$

and total number of vitamins v_3 (*in* mg) equal to $10x_1 + 100x_2$

Again since the maximum daily requirement of vitamin v_1 is 1 mg of vitamin v_2 is 50 mg and of vitamin v_3 is 100 mg therefore, we must have

$$x_1 + x_2 \geq 1$$

$$100x_1 + 10x_2 \geq 50$$

$$10x_1 + 100x_2 \geq 10$$

Also, since x_1 and x_2 are either positive or zero, we have

$$x_1 \geq 0 \text{ and } x_2 \geq 0$$

Hence this diet problem can be put in the following mathematical form:

Find two real numbers x_1 and x_2 such that

$$x_1 + x_2 \geq 1$$

$$100x_1 + 10x_2 \geq 50$$

$$10x_1 + 100x_2 \geq 10$$

$$x_1, x_2 \geq 0$$

and for which the expression (objectively function)

$$Z = x_1 + 1.5x_2$$

may be a minimum (least).

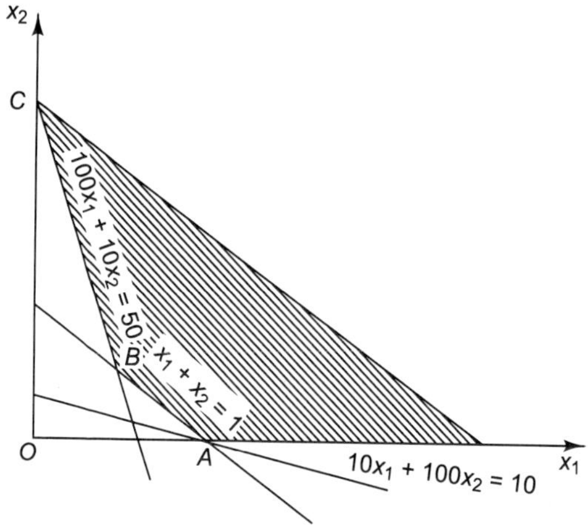

Fig. 7

Graphical Solution: Let us consider a set of rectangular cartesian axis Ox_1x_2 in the plane. Each point has co-ordinates of the type (x_1, x_2). Clearly, any point (x_1, x_2) satisfying the condition $x_1 \geq 0$ and $x_2 \geq 0$, lies in the first quadrant only. Thus we restrict out search for the number pair (x_1, x_2) in the first quadrant only. Also since

$$x_1 + x_2 \geq 1$$

$$100x_1 + 10x_2 \geq 50$$

and $$10x_1 + 100x_2 \geq 10$$

the desired point (x_1, x_2) must be somewhere in the shaded region, which is unbounded.

Since our search is for such a number pair (x_1, x_2) which gives a minimum value of Z, the desired point will be at any one of the three vertices of the shade region.

The co-ordinates of three vertices are

$$A = (1, 0); B = (0.4, 0.6) \text{ and } C = (0, 5)$$

Values of the objective function at these three vertices are

$$Z(A) = 1, Z(B) = 1.3 \text{ and } Z(C) = 7.5$$

Obviously minimum value of Z is 1 which occurs at $A = (1, 0)$

Hence the solutions to this diet problem is

$$x_1 = 1, x_2 = 0; \text{Min } Z = 1.$$

10. Solve graphically the following L.P.P.

Maximize $Z = 3x_1 + 2x_2$

Subject to constraints

$$-2x_1 + x_2 \leq 1$$

$$x_1 \leq 2$$

$$x_1 + x_2 \leq 3$$

$$x_1, x_2 \geq 0$$

Solution:

Graphical Solution: Let us consider a set of rectangular cartesian axis Ox_1x_2 in the plane. Each point has co-ordinates of the type (x_1, x_2) and conversely. It is clear that any point (x_1, x_2) which satisfies the conditions $x_1 \geq 0, x_2 \geq 0$ lies in the first quadrant only. Thus our search for the desired number pair (x_1, x_2) will be restricted to the first quadrant only. Also since

$$-2x_1 + x_2 \leq 1$$

$$x_1 \leq 2$$

and $$x_1 + x_2 \leq 3$$

the required point (x_1, x_2) lies somewhere in the region $OABNM$ which is shown shaded in the figure. The co-ordinates of five vertices of region are $O = (0, 0)$ $A = (2, 0)$, $B = (2, 1)$.

$$N = \left(\frac{2}{3}, \frac{7}{3}\right) \text{ and } M = (0, 1)$$

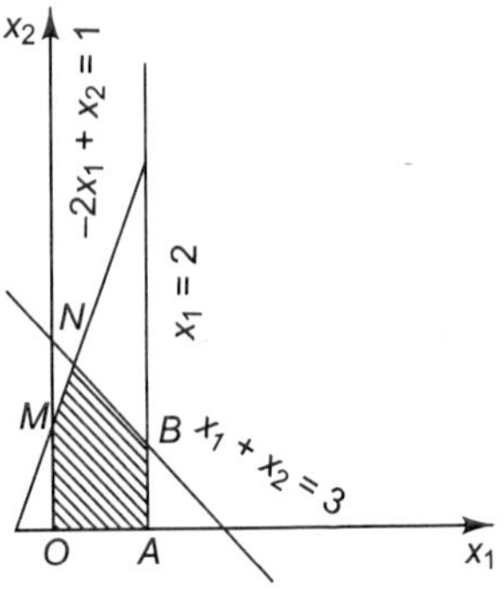

Fig. 8

Let $Z = f(x_1, x_2) = 3x_1 + 2x_2$

$Z(O) = f(0, 0) = 3.0 + 2.0 = 0$

$Z(A) = f(2, 0) = 3.2 + 2.0 = 6$

$Z(B) = f(2, 1) = 3.2 + 2.1 = 8$

$$Z(N) = f\left(\frac{2}{3}, \frac{7}{3}\right) = 3 \times \frac{2}{3} + 2 \times \frac{7}{3} = 6.7$$

$Z(M) = f(0, 1) = 3 \times 0 + 2 \times 1 = 2$

Thus the maximum value of Z is 8 and this value occurs at $B = (2, 1)$

Hence, the solutions $x_2 = 2$, $x_2 = 1$, $Z = 8$

11. Solve graphically the following L.P.P.:

Max. $Z = 3x_1 + 2x_2$

Subject to $$x_1 - x_2 \leq 1$$

$$x_1 + x_2 \geq 3$$

$$x_1, x_2 \geq 0$$

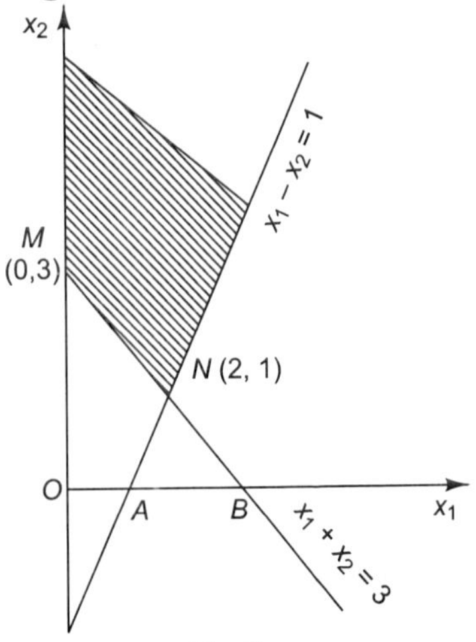

Fig. 9

Let us consider a set of rectangular axes Ox_1, x_2. Any point (x_1, x_2) satisfying the condition $x_1 \geq 0$ and $x_2 \geq 0$ lies in the first quadrant only. Also, since $x_1 - x_2 \leq 1$ and $x_1 + x_2 \geq 3$, the required point must be somewhere in the shaded portion. Hence the shaded portion is unbounded.

Here the shaded region (solution space) is unbounded.

The two vertices of the region in the finite plane are

$$N = (2, 1) \qquad \text{and } M = (0, 3)$$

The values of the objective function at these vertices are

$$Z(M) = 6 \qquad \text{and } Z(N) = 8$$

But the exist points in the convex region for which the value of the objective function is more than 8. For instance, the point (16, 10) lies in the region and function-value at this point is 50 which is more than 8. Hence, the maximum value of Z occurs at the point at infinity only and thus the problem has an unbounded solution.

12. Solve graphically the following L.P.P.

Maximize $\quad Z = x_1 + x_2$

Subject to the constraints:

$$x_1 + x_2 \leq 1$$
$$-3x_1 + x_2 \geq 3$$
$$x_1, x_2 \geq 0$$

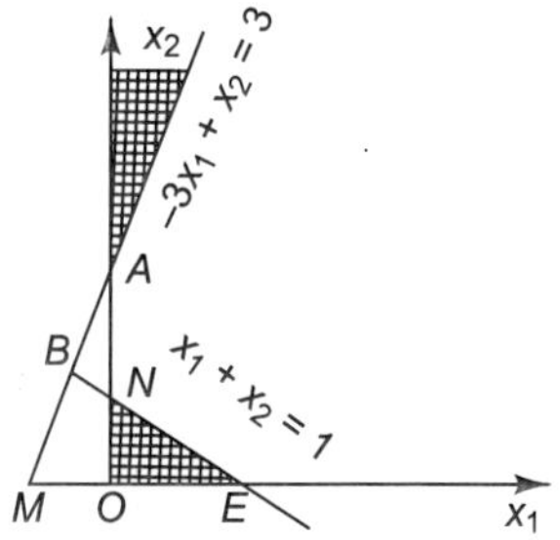

Fig. 10

Graphical Solution: Let us consider a set of rectangular cartesians axis Ox_1x_2 in the plane. Clearly the desired point (x_1, x_2) lies in the first quadrant only. In this case the two half planes.

$$x_1 + x_2 \leq 1$$

and $\quad -3x_1 + x_2 \geq 3$

do not intersect and thus have no point in common.

The region common to the two half planes and the first quadrant are shown as shaded portion.

Since there is no point (x_1, x_2) which can be in both the regions these exists no solution to the given L.P.P.

13. A toy Company manufactures two types of doll, a basic version doll A and a deluxe version doll B. Each doll of type B takes twice as long to produce as one type A, and the company would have time to make a maximum of 2,000 per day if it produces only the basic version. The supply of plastic is sufficient to produce 1500 dolls per day both A and B combined, the deluxe version requires a fancy dress of which there are only 600 per day available. If the Company makes of Rs. 3.00 and Rs. 5.00 per doll respectively, on doll A and B; how many of each should be produced per day in order to maximize profit?

Solution: Let x_1 dolls of type A and x_2 dolls of type B be produced per day.

Therefore, total profit $\quad Z = 3x_1 + 5x_2$

Total time per day consumed to prepare x_1 and x_2 dolls of type A and B is $x_1 + 2x_2$ which should be less than 2,000, Therefore,

$$x_1 + 2x_2 \leq 2,000$$

Since plastic is available to produce 1500 dolls only therefore,

$$x_1 + x_2 \leq 1500$$

Also fancy dress is available for 600 dolls per day only

$$\therefore x_2 \leq 600$$

Hence the linear programming problem is as follows.

$$\text{Max. } Z = 3x_1 + 5x_2$$

Subject to the constraints

$$x_1 + 2x_2 \leq 2000$$

$$x_1 + x_2 \leq 1500$$

$$x_2 \leq 600$$

and $\qquad x_1 \geq 0, x_2 \geq 0$

Now, first consider the constraints as equalities

$$x_1 + 2x_2 = 2000$$

$$x_1 + x_2 = 1500$$

$$x_2 = 600$$

The graphical representation of three lines are shown in the figure.

The permissible region is shown shaded portions. Obtain the co-ordinate of vertices.

By solving $\qquad x_1 + 2x_2 = 2000$

and $\qquad x_1 + x_2 = 1500$

we get the co-ordinates of P.

P is $\qquad (1000, 500)$

Put the co-ordinates of P in

$$Z = 3x_1 + 5x_2$$

Then Max. Z = Rs. $\{3 \times 1000 + 5 \times 500\}$ = Rs. 5500.

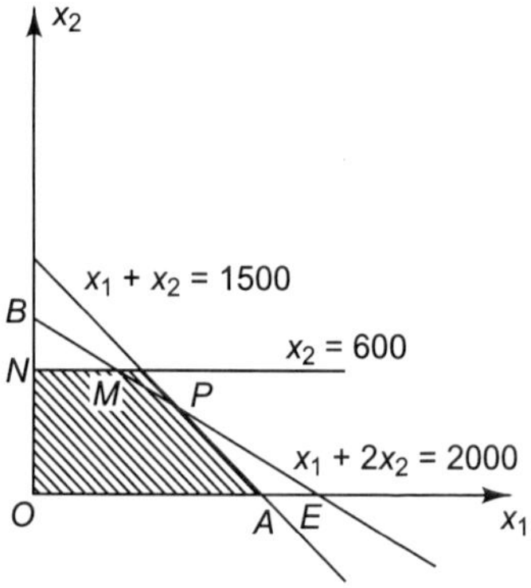

Fig. 11

14. Solve the problem graphically.

$$\text{Max. } Z = 6x_1 - 2x_2$$

Subject to $\qquad 2x_1 - x_2 \leq 2$

$$x_1 \leq 3$$

$$x_1, x_2 \geq 0$$

Solution: The constraints are put as equalities

$$2x_1 - x_2 = 2 \qquad \text{...(1)}$$

$$x_1 = 3 \qquad \text{...(2)}$$

Solving (1) and (2), we get $x_1 = 3, x_2 = 4$

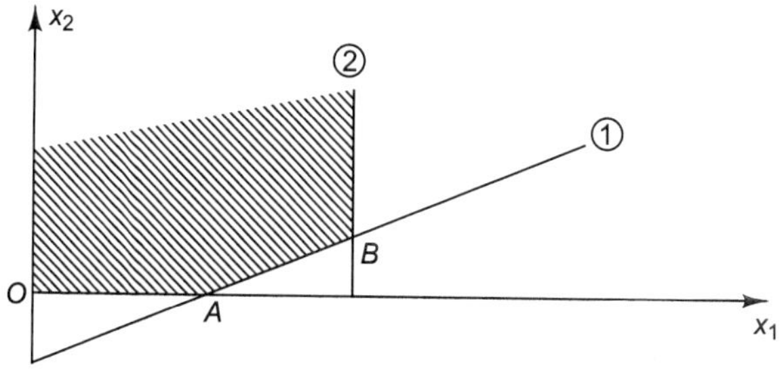

Fig. 12

Here, it is to be noted that the permissible region is unbounded but finite maximum exists.

Thus always unbounded region does not mean unbounded solution.

The optimum solution is attained at $B(3, 4)$

$$Z\text{max.} = 6 \times 3 - 2 \times 4$$
$$= 10$$

15. Solve graphically the following L.P.P.

Maximize $\quad Z = x_1 + \frac{1}{2} x_2$

Subject to the constraints

$$3x_1 + 2x_2 \le 12$$
$$5x_1 \le 10$$
$$x_1 + x_2 \ge 8$$
$$-x_1 + x_2 \ge 4$$

and $\quad x_1, x_2 \ge 0$

Solution: Convert the constraints as follows equations

$$3x_1 + 2x_2 = 12 \quad ...(1)$$
$$5x_1 = 10 \quad ...(2)$$
$$x_1 + x_2 = 8 \quad ...(3)$$
$$-x_1 + x_2 = 4 \quad ...(4)$$

The adjoining figure depicts the permissible region (shaded). The two shaded areas indicate non-overlapping regions that can be considered as feasible solution area in the sense that they satisfy some subsets of the constraints. However, since we can not find any (x_1, x_2) to be in both these two shaded regions, the present problem can not be solved by using usual graphical method of solving L.P.P.)

16. Solve the problem graphically

Maximum $Z = 3x_1 + 4x_2$

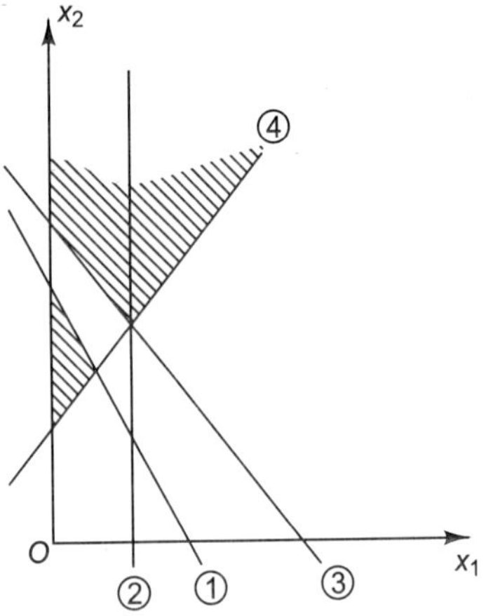

Fig. 13

Subject to $5x_1 + 4x_2 \leq 200$

$3x_1 + 5x_2 \leq 150$

$5x_1 + 4x_2 \geq 100$

$8x_1 + 4x_2 \geq 80$

and $x_1, x_2 \geq 0$

Solution: Converting the inequalities into equations

$$5x_1 + 4x_2 = 200 \quad ...(1)$$

$$3x_1 + 5x_2 = 150 \quad ...(2)$$

$$5x_1 + 4x_2 = 100 \quad ...(3)$$

$$8x_1 + 4x_2 = 80 \quad ...(4)$$

The optimal solution of the problem occurs at the point C which is the intersection of the line.

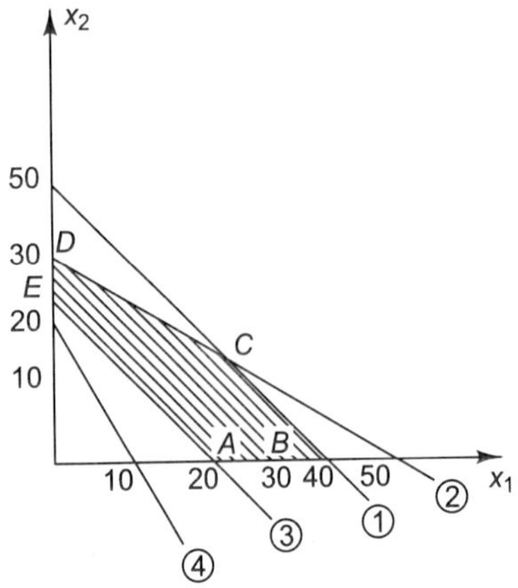

Fig. 14

$$3x_1 + 5x_2 = 150 \quad \text{... (2)}$$

and $$5x_1 + 4x_2 = 200 \quad \text{... (1)}$$

It is to be seen that the constraint $8x_1 + 4x_2 \geq 80$ imposes no extra restriction and hence does not affect the solution.

Such a constraint is called a Redundant or superfluous constraint .

Solving (1) and (2), we get

$$x_1 = 30.7, x_2 = 11.5$$

$$\begin{aligned}\text{Maximum } Z &= 3x_1 + 4x_2 \\ &= 3\,(30.7) + 4\,(11.5) \\ &= 92.1 + 46.0 \\ &= 138.1\end{aligned}$$

17. Solve by graphical method

$$\text{Max. } Z = x_1 + x_2$$

Subject to
$$2x_1 + 3x_2 \leq 22$$
$$2x_1 + x_2 \leq 14$$
$$x_1 - x_2 \leq 4$$
$$3x_1 - 2x_2 \leq - 6$$
$$x_1, x_2 \geq 0$$

Solution: Putting the constraints as equations

$$2x_1 + 3x_2 = 22 \quad \text{...(1)}$$
$$2x_1 + x_2 = 14 \quad \text{...(2)}$$
$$3x_1 - 2x_2 = - 6 \quad \text{...(3)}$$
$$x_1 - x_2 = 4 \quad \text{...(4)}$$

The graphical representation is shown in the figure. Points of intersection are $A(4, 0)$, $b(6, 2)$, $C(5, 4)$ $d(2, 6)$ and $E(0, 3)$.

$$Z = x_1 + x_2$$
$$Z_A = 4$$
$$Z_B = 6 + 2 = 8$$
$$Z_C = 5 + 4 = 9$$
$$Z_D = 2 + 6 = 8$$
$$Z_E = 0 + 3 = 3$$

Clearly the required solution is $x_1 = 5, x_2 = 4$

$$Z \text{ max.} = 9.$$

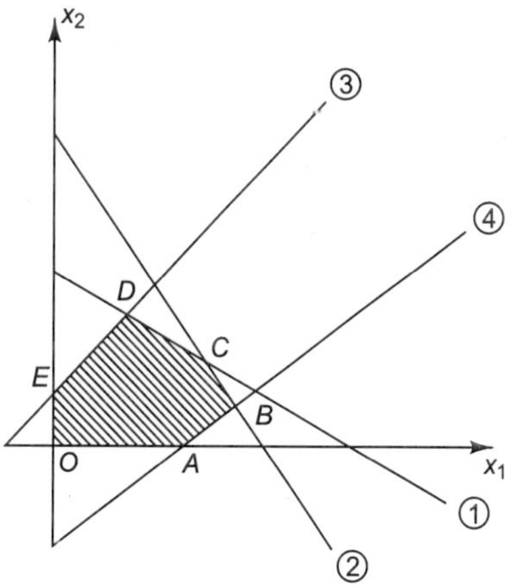

Fig. 15

18. Solve the L.P.P. graphically

$$\text{Max. } Z = 8000x_1 + 7000x_2$$

Subject to
$$3x_1 + x_2 \leq 66$$
$$x_1 + x_2 \leq 45$$
$$x_1 \leq 20$$
$$x_2 \leq 40$$
and $x_1, x_2, \leq 0$ [Meerut '73]

Solution: Converting the constraints into equations

$$3x_1 + x_2 = 66 \quad ...(1)$$
$$x_1 + x_2 = 45 \quad ...(2)$$
$$x_1 = 20 \quad ...(3)$$
$$x_2 = 40 \quad ...(4)$$

Shaded portion is permissible region.

Points A, B, N, M, C are $A(20, 0)$, $B(20, 6)$, $N(10.5, 34.5)$, $M(5, 40)$, $C(0, 40)$

Let $Z = f(x_1, x_2) = 8000x_1 + 7000x_2$

Then $Z_A = f(20, 0) = 160000$

$Z_B = f(20, 0) = 202000$

$Z_N = 325500$

$Z_M = 320000$

$Z_C = 280000$

Z max. $= 325500$

for $x_1 = 10.5$

$x_2 = 34.5$

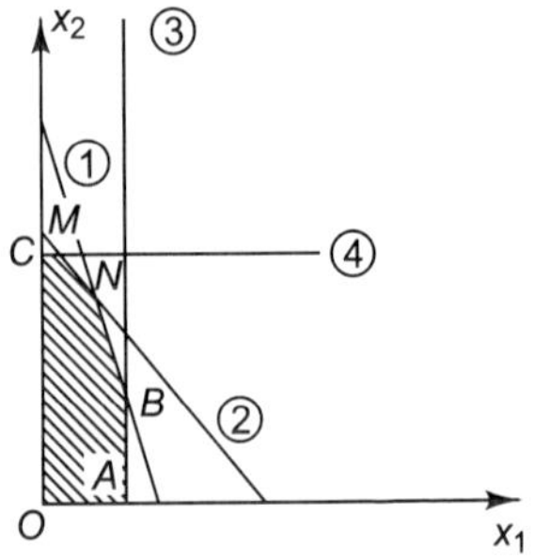

Fig. 16

19. Solve graphically,

$$\text{Max. } Z = -x_1 + 2x_2$$

Subject to
$$-x_1 + x_2 \leq 1$$
$$-x_1 + 2x_2 \leq 4$$
$$x_1, x_2 \geq 0$$

Solution: Putting the inequalities as equations

$$-x_1 + x_2 = 1 \quad ...(1)$$
$$-x_1 + 2x_2 = 4 \quad ...(2)$$

Here, the line of objective function coincides with the edge of the permissible region of solutions (A, ∞). Thus any point (x_1, x_2) lying on this edge which extends to infinity gives $Z = 4$ and is therefore, an optimal solution.

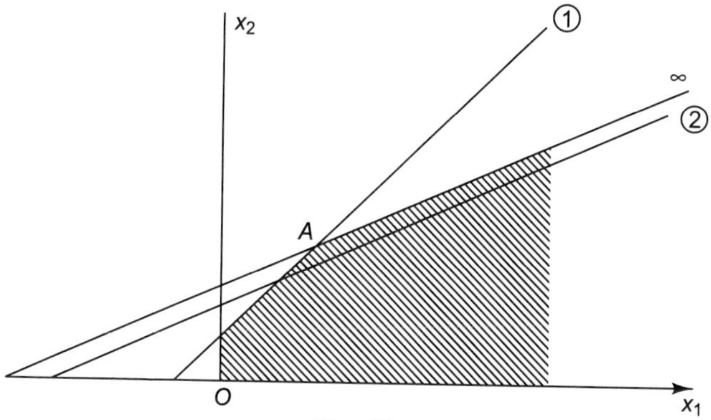

Fig. 17

20. Solve the L.P.P. by graphical method,

$$\text{Max. } Z = 3x_1 - 2x_2$$

Subject to $\quad x_1 + x_2 \le 1$

$\quad 2x_1 + 2x_2 \ge 4$

and $x_1, x_2 \ge 0$

Solution: Putting the constraints as qualities

$x_1 + x_2 = 1 \qquad ...(1)$

$2x_1 + 2x_2 = 4 \qquad ...(2)$

The figure shows that there is no point (x_1, x_2) which satisfies both the constraints.

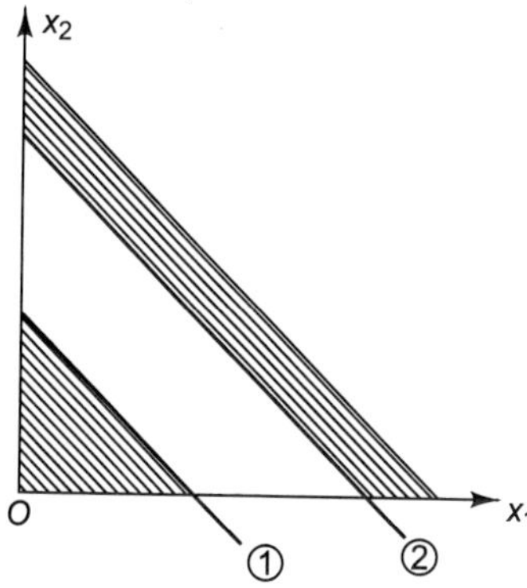

Fig. 18

Here, the problem has no solution because the constraints are inconsistent.

21. Solve $\quad \text{Max. } Z = 5x_1 + 3x_2 \qquad ...(1)$

Subject to $\quad 3x_1 + 5x_2 = 15 \qquad ...(2)$

$\quad 5x_1 + 2x_2 = 10$

by graphical method

Since there is only a single solution point $M\left(\frac{20}{19}, \frac{45}{19}\right)$ there is nothing to be maximized. Hence a problem of this kind is of no interest. Such problem can arise only when the number of equations in the constraints is at least equal to the number of variables. If the solution is feasible, it is optimal. If it is not feasible, the problem has no solution.

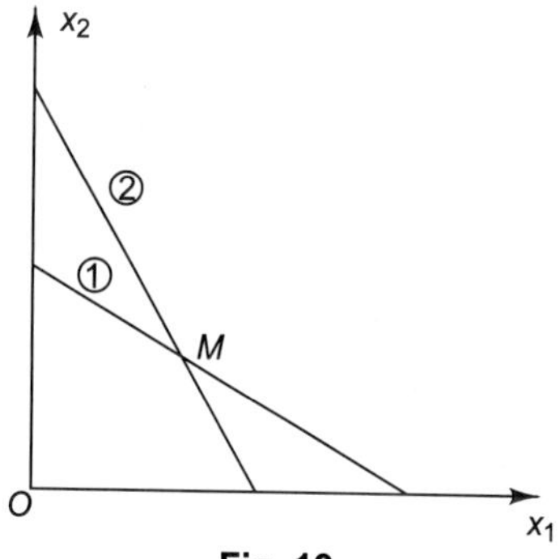

Fig. 19

22. Solve graphically the following linear programming problem.

$$\text{Min } Z = 3x_1 + 5x_2$$

s.t.

$$-3x_1 + 4x_2 \le 12$$
$$x_1 \le 4$$
$$2x_1 - x_2 \ge -2$$
$$x_2 \ge 2$$
$$2x_1 + 3x_2 \ge 12,$$
$$x_1, x_2 \ge 0$$

Solution: Putting the constraints as the equalities, we have

$$-3x_1 + 4x_2 = 12 \quad ...(1)$$
$$x_1 = 4 \quad ...(2)$$
$$2x_1 - x_2 = -2 \quad ...(3)$$
$$x_2 = 2 \quad ...(4)$$
$$2x_1 + 3x_2 = 12 \quad ...(5)$$

The feasible region is shown as shaded portion.

The co-ordinates of the point A, B, C, D and E can be obtained by solving the respective equations. They are $A(3, 2)$.

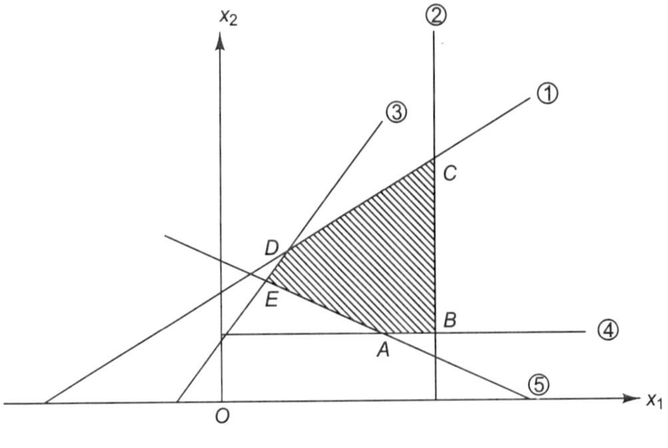

Fig. 20

$$B(4, 2); \qquad C(4, 6)$$

$$D\left(\frac{4}{5}, \frac{18}{5}\right) \qquad E\left(\frac{3}{4}, \frac{14}{4}\right)$$

Let $\quad Z = f(x_1, x_2) = 3x_1 + 5x_2$

$$Z_A = f(3, 2) = 3 \times 3 + 5 \times 2 = 19$$

$$Z_B = f(4, 2) = 22$$

$$Z_C = f(4, 6) = 42$$

$$Z_D = f\left(\frac{4}{5}, \frac{18}{5}\right) = 20.4$$

$$Z_E = f\left(\frac{3}{4}, \frac{14}{4}\right) = 19.75$$

Z min $= 19$ for $x_1 = 3, x_2 = 2$

23. The graphical method (A Three Dimensional Case)

Solving graphically L.P.P.

$$\text{Max. } Z = 3x_1 + 2x_2 + x_3$$

s.t. $5x_1 + 6x_2 + 10x_3 \le 50$ ...(2)

$5x_1 + 4x_2 \le 20$ ...(3)

$15x_1 + 4x_2 \le 30$ ...(4)

and $x_1, x_2, x_3 \ge 0$ ...(5)

Solution: Using equality signs the graph has been drawn in three dimensional space. The space *OABCD EFG* in the first octant i.e. the space bounded by the three co-ordinate, planes $x_1 = 0, x_2 = 0, x_3 = 0$ and three other planes *ABGF*, *BCDG* and *DEFG* is the required feasible space. The boundaries of the feasible space are shaded in the figure. After solving the equations of the boundaries.

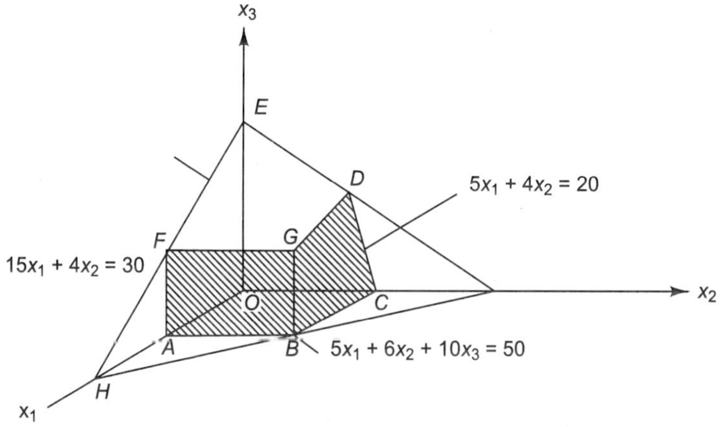

Fig. 21

$$5x_1 + 6x_2 + 10x_3 = 50$$

$$5x_1 + 4x_2 = 20$$

$$15x_1 + 4x_2 = 30$$

$x_1 = 0, x_2 = 0, x_3 = 0$, taken three at a time we get the points

$$A(2, 0, 0), B\left(1, \frac{15}{4}, 0\right), C(0, 5, 0), D(0, 5, 0)$$

$$E(0, 0, 5), P(2, 0, 4), G\left(1, \frac{15}{4}, \frac{9}{4}\right)$$

Let $\quad Z = f(x_1, x_2, x_3) = 3x_1 + 2x_2 + x_3$

$\therefore \quad Z_A = f(2, 0, 0) = 3 \times 2 + 2 \times 0 + 0 = 6$

$$Z_B = f\left(1, \frac{15}{4}, 0\right) = 3 \times 1 + 2 \times \frac{15}{4} + 0 = \frac{21}{2}$$

$$Z_C = f(0, 5, 0) = 3 \times 0 + 2 \times 5 + 0 = 10$$

$$Z_D = f(0, 5, 2) = 3 \times 0 + 2 \times 5 + 2 = 12$$

$$Z_E = f(0, 0, 5) = 3 \times 0 + 2 \times 0 + 5 = 5$$

$$Z_F = f(2, 0, 4) = 3 \times 2 + 2 \times 0 + 4 = 10$$

$$Z_G = f\left(1, \frac{15}{4}, \frac{9}{4}\right) = 3 \times 1 + 2 \times \frac{15}{4} + \frac{9}{4} = \frac{51}{4}$$

The maximum occurs at

$$x_1 = 1, x_2 = \frac{15}{4}, x_3 = \frac{9}{4} \text{ and}$$

Maximum value of Z is $\frac{51}{4}$.

24. Every would like to decide on the constituents of a diet which will satisfy his daily needs of proteins, fats and carbohydrates at the minimum cost. Choices from three different type of foods can be made. The yields per unit of the food are given by

Food type	*Yields per unit*			*Cost per Unit Rs.*
	Proteins	*Fats*	*Carbo-hydrates*	
Bread	4	1	2	2
Butter	3	2	1	9
Milk	3	2	1	1
Daily requirement	5	5	3	

Obtain Linear Programming formulation of this problem.

Formulation; Let x_1, x_2, x_3 be the numbers of units used for bread, butter and milk respectively. Thus the minimum daily requirements of the person is satisfied if

$$4x_1 + 3x_2 + 3x_3 \geq 5$$

$$x_1 + 2x_2 + 2x_3 \geq 2$$

$$2x_1 + 1x_2 + 1x_3 \geq 3$$

where $x_1 \geq 0, x_2 \geq 0, x_3 \geq 0$.

The objective function is

Min. $Z = 2x_1 + 9x_2 + 1x_3$

25. Solve graphically the following L.P.P.

Max. $Z = 5x + 3y$

Subject to $3x + 5y \leq 15$

$5x + 2y \leq 10$

and $x, y \geq 0$

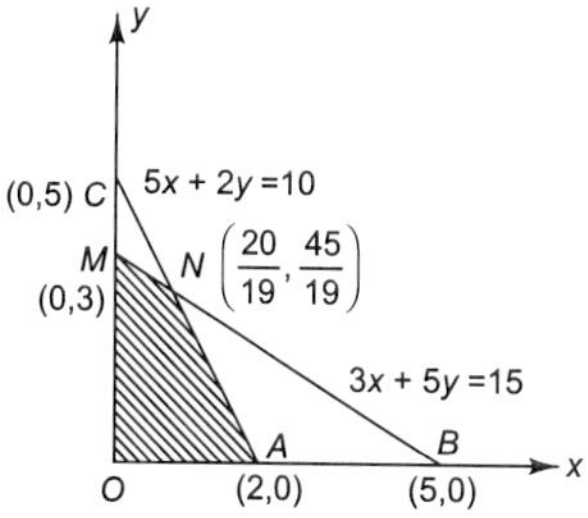

Fig. 22

[R.U.B.Sc. '86, 87]

The feasible region is shown by the shaded portion in the graphical representation, co-ordinates of the points are shown in the figure

Let $Z = f(x, y) = 5x + 3y$

$$Z_O = f(0, 0) = 0$$

$$Z_A = f(2, 0) = 5 \times 2 + 3 \times 0 = 10$$

$$Z_N = f\left(\frac{20}{19}, \frac{45}{19}\right) = 5 \times \frac{20}{19} + 3 \times \frac{45}{19} = 12\frac{7}{19}$$

$$Z_M = f(0, 3) = 5 \times 0 + 3 \times 3 = 9$$

$$\therefore \quad Z_{Max.} = 12\frac{7}{19} \text{ for } x = \frac{20}{19}, y = \frac{45}{19}$$

26. Solve the following L.P.P. by graphical method

Max. $Z = 5x + 7y$

Subject to $x + y \leq 4$

$3x + 8y \leq 24$

$10x + 7y \leq 35$

$x, y \geq 0$

[Raj T D C 76 Non-collegiate, Meerut L.P. 70, 76, 77, 86, 87, 89]

Permissible region is shown by shaded region.

The co-ordinates of the extremities of the region are as below:

$M = (3.5, 0)$

$E = (7/3, 5/3)$

$N = (8/5, 12/5)$

$A = (0, 3)$

Let $Z = f(x, y) = 5x + 7y$

$Z_O = f(0, 0) = 0$

$Z_M = f(3.5, 0) = 5 \times 3.5 + 7 \times 0 = 17.5$

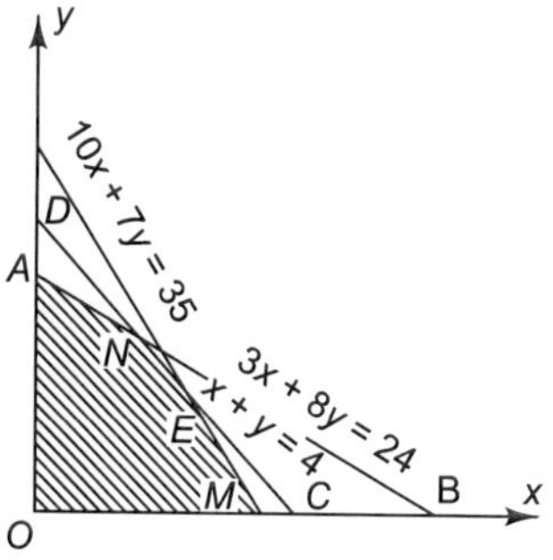

Fig. 23

$$Z_N = f(8/5, 12/5) = 5 \times \frac{8}{5} + 7 \times \frac{12}{5} = 24.8$$

$$Z_A = f(0, 3) = 5 \times 0 + 7 \times 3 = 21$$

$$Z_{\text{Max.}} = 24.8 \text{ for } x = \frac{8}{5}, y = \frac{12}{5}$$

27. Solve the following L.P.P. by graphical method.

Max. $Z = 3x_1 + 2x_2$

$x_1 - x_2 \leq 1$

$x_1 + x_2 \geq 3$

$x_1, x_2 \geq 0$

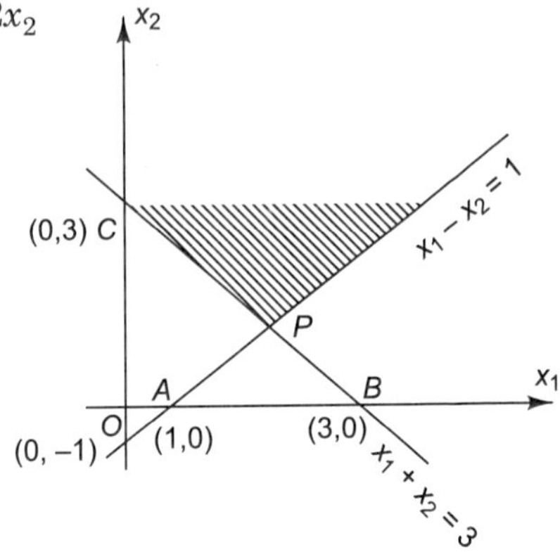

Fig. 24

The graphical representation shows that $Z = 3x_1 + 2x_2$ can be increased indefinitely.

Therefore, Z is unbounded.

28. Solve the following L.P.P. by graphical method.

Max. $Z = x + y$

s.t. $x + 2y \leq 20$

$x + y \leq 15$

$y \leq 6$

$x, y \geq 0$

The permissible region is shown by the shaded portion. The co-ordinates of A, P, C, D are (15, 0), (10, 5), (8, 6) and (0, 6) respectively.

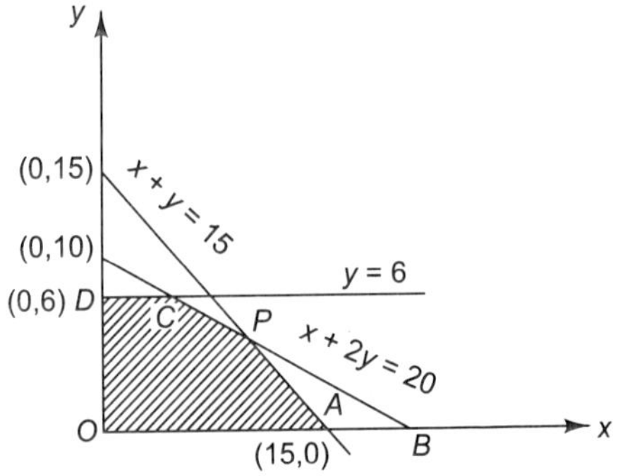

Fig. 25

Solving $x + y = 15$

and $x + 2y = 20$

we get $x = 10, y = 5$

Hence P is (10, 5)

Again solving $y = 6$

and $x + 2y = 20$

We get $x = 8, y = 6$

$\therefore$ C is (8, 6)

Let $Z = f(x, y) = x + y$

$Z_O = 0$

$Z_A = f(15, 0) = 15 + 0 = 15$

$Z_P = f(10, 5) = 10 + 5 = 15$

$Z_C = f(8, 6) = 8 + 6 = 14$

$Z_D = f(0, 6) = 0 + 6 = 6$

$Z_{\text{Max}} = 15$

Observation: Maximum value of Z is attained at two points A and P. Hence the solution is not unique. Any point an AP will give the optimal solution.

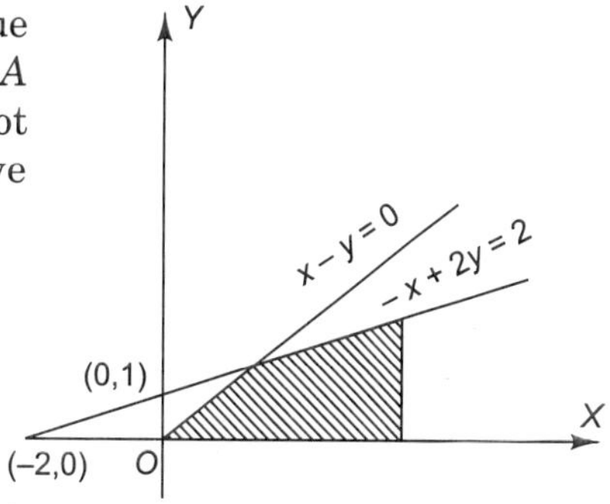

Fig. 26

29. Solve graphically by

$$\text{Max. } Z = 3x + 4y$$

s.t.
$$x - y \geq 0$$
$$-x + 2y \leq 2$$
$$x, y \geq 0$$

From figure it is obvious that Z can be increased indefinitely. Hence Z is unbounded.

30. Solve the following L.P.P. graphically:

$$\text{Max. } Z = 2x + 3y$$

s.t.
$$x - y \geq 0$$
$$-3x + y \geq 3$$
$$x, y \geq 0$$

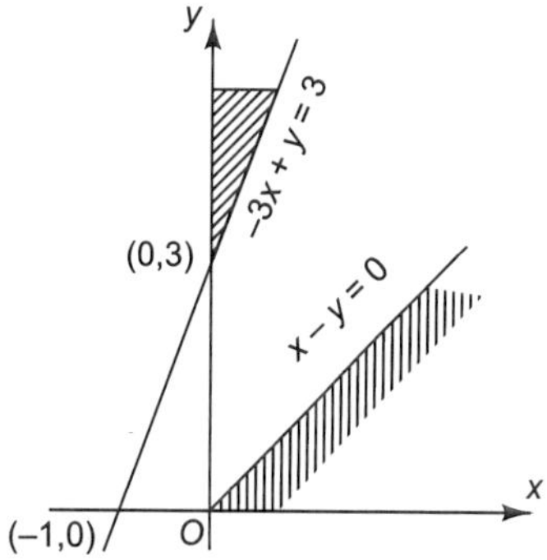

Fig. 27

The required feasible region is shown in the figure by shaded region. From the figure it is obvious that the solution is not consistent. Hence maximum value of Z can not be attained.

31. Solve graphically the following L.P.P.:

$$\text{Min. } Z = -x_1 + 2x_2$$

s.t.
$$-x_1 + 3x_2 \leq 10$$
$$x_1 + x_2 \leq 6$$
$$x_1 - x_2 \leq 2$$
$$x_1, x_2 \geq 0$$

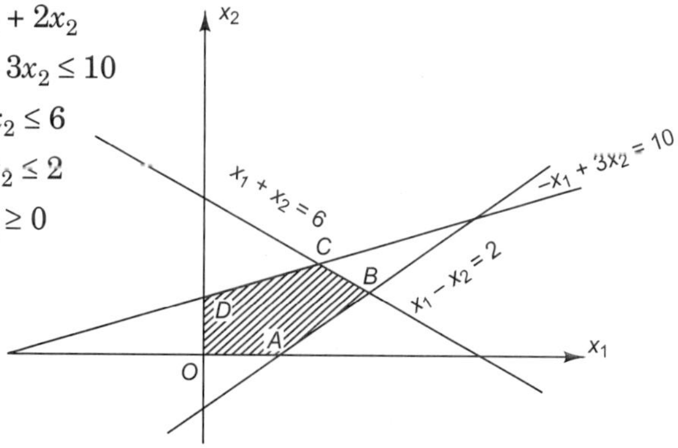

Fig. 28

The shaded portion is the feasible region. A is $(2, 0)$, D is $\left(0, \frac{10}{3}\right)$

To get the co-ordinates of B

Solve $x_1 + x_2 = 6$, and $x_1 - x_2 = 2$

Then B is $(4, 2)$

To get the Co-ordinates of C
solve $x_1 + x_2 = 6$ and $-x_1 + 3x_2 = 10$.
Then C is (2, 4)
Let $Z = f(x_1, x_2) = -x_1 + 2x_2$

$$Z_A = f(2, 0) = -2$$

$$Z_B = f(4, 2) = -4 + 2 \times 2 = 0$$

$$Z_C = f(2, 4) = -2 + 4 \times 2 = 6$$

$$Z_D = f(0, 10/3) = 0 + 2 \times \frac{10}{3} = \frac{20}{3}$$

$\therefore$ Zmin. $= -2$ for $x_1 = 2, x_2 = 0$

32. Solve graphically

Min. $Z = 2x_1 + 3x_2$

s.t.
$$x_1 + x_2 \le 4$$
$$6x_1 + 2x_2 \ge 8$$
$$x_1 + 5x_2 \ge 4$$
$$x_1 \le 3$$
$$x_2 \le 3$$
$$x_1, x_2 \ge 0$$

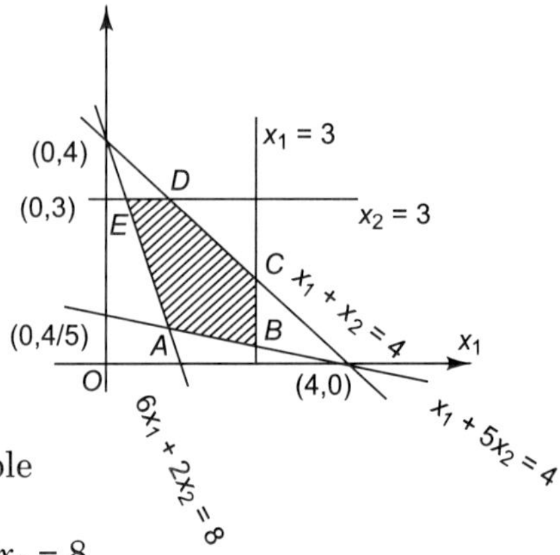

Fig. 29

The shaded region is the feasible region.

To get A we solve, $6x_1 + 2x_2 = 8$
and $x_1 + 5x_2 = 4$ $\therefore A$ is (8/7, 4/7)

To get B we solve, $x_1 + 5x_2 = 4$
and $x_1 = 3$ $\therefore B$ is (3, 1/5)

To get C we solve, $x_1 + x_2 = 4$
and $x_1 = 3$ $\therefore C$ is (3, 1)

To get D we solve, $x_1 + x_2 = 4$
and $x_2 = 3$ $\therefore D$ is (1, 3)

To get E we solve, $6x_1 + 2x_2 = 8$
and $x_2 = 3$ $\therefore E$ is (1/3, 3)

Let $Z = f(x_1, x_2) = 2x_1 + 3x_2$

$$Z_A = f(8/7, 4/7) = 4$$

$$Z_B = f(3, 1/5) = 33/5$$

$$Z_C = f(3, 1) = 9$$

$$Z_D = f(1, 3) = 11$$

$$Z = f\left(\frac{1}{3}, 3\right) = 29/3$$

$\therefore$ Zmin. $= 4$ for $x_1 = 8/7, x_2 = 4/7$.

33. Determine $x \geq 0, y \geq 0$, so as to maximize Z = 2x + 3y

s.t. $x + y \leq 30$

$y \leq 12$

$x \leq 20$

$y \leq 3$

$x, y \leq 0$

by graphical method

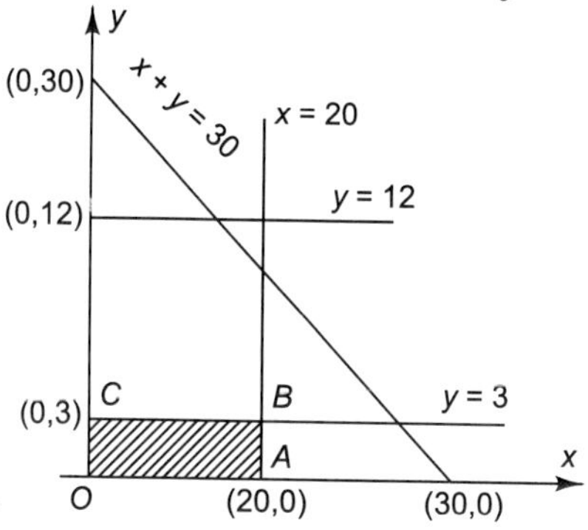

Fig. 30

The shaded region is the feasible region. The extremities of the feasible region are

$O(0, 0), A = (20, 0), B(20, 3), C(0, 3)$

Let $Z = f(x, y) = 2x + 3y$

$$Z_A = f(20, 0) = 40$$

$$Z_B = f(20, 3) = 49$$

$$Z_C = f(0, 3) = 9$$

Z max = 49 for $x = 20, y = 3$

34. Solve graphically:

$$\text{Max. } Z = 3x_1 + 2x_2$$

s.t. $2x_1 - x_2 \geq 2$

$x_1 + 2x_2 \leq 8$

$x_1, x_2 \geq 0$

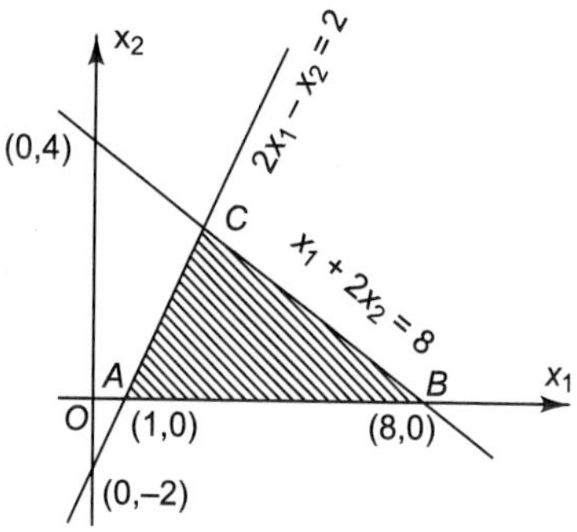

Fig. 31

The shaded region is the feasible region. Clearly A is (1, 0) B is (8, 0). To get co-ordinates of C

We solve $2x_1 - x_2 = 2$

and $x_1 + 2x_2 = 8$

$\therefore$ C is (12/5, 14/5)

Let

$$Z = f(x_1, x_2) = 3x_1 + 2x_2$$

$$Z_A = f(1, 0) = 3$$

$$Z_B = f(8, 0) = 24$$

$$Z_C = f(12/5, 14/5) = 64/5$$

$\therefore$ Z max = 24 for $x_1 = 8, x_2 = 0$

35. Solve graphically:

Min. $Z = 4x_1 + 2x_2$

s.t. $x_1 + 2x_2 \geq 2$

$3x_1 + x_2 \geq 3$

$4x_1 + 3x_2 \geq 6$

$x_1, x_2 \geq 0$ [R.U.B.Sc. 88]

Feasible region is shown by shaded portion.

Solving $x_1 + 2x_2 = 2$ and $4x_1 + 3x_2 = 6$.

We get the co-ordinates of B

$\therefore$ B is (6/5, 2/5)

Similarly solving $3x_1 + x_2 = 3$

and $4x_1 + 3x_2 = 6$,

we get the co-ordinate of C.

$\therefore$ C is (3/5, 6/5).

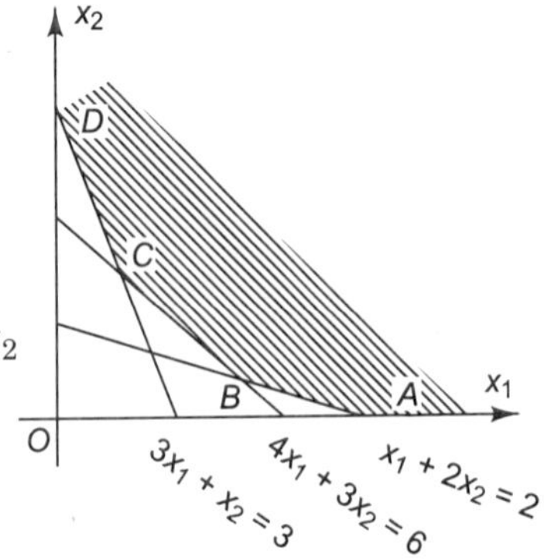

Fig. 32

Let $Z = f(x_1, x_2) = 4x_1 + 2x_2$

$Z_A = f(2, 0) = 8$

$Z_B = f(6/5, 2/5) = 28/5$

$Z_C = f(3/5, 6/5) = 24/5$

$Z_D = f(0, 3) = 6$

$\therefore$ $Z \text{ min.} = \dfrac{24}{5}$ for $x_1 = 3/5, x_2 = 6/5$

36. A baker has 90, 80 and 50 units of gradients A, B and C respectively. A loaf of bread requires 2, 1 and 1 units of A, B and C respectively; and a cake requires 1, 2 and 1 units of A, B and C respectively. If a loaf of bread sells for a cake for a profit of Rs. 0.30 and a cake for Rs. 0.40, how many of each should be baker so that his profit is maximum.

Solution: Let the number of breads be x_1 and than of cake be x_2 we make the following table on the basis of the above problem.

Ingradient	*Product*		*Requirement*
	Bread (x_1)	*Cake* (x_2)	
A	2	1	90
B	1	2	80
C	1	1	50
Profit	Rs. 0.30	Rs. 0.40	

The mathematical formulation of the above problem is as follows:

$$\text{Max. } Z = 30x_1 + 40x_2$$

subject to

$$2x_1 + x_2 \geq 90$$

$$x_1 + 2x_2 \leq 80$$

$$x_1 + x_2 \leq 50$$

$$x_1, x_2 \geq 0$$

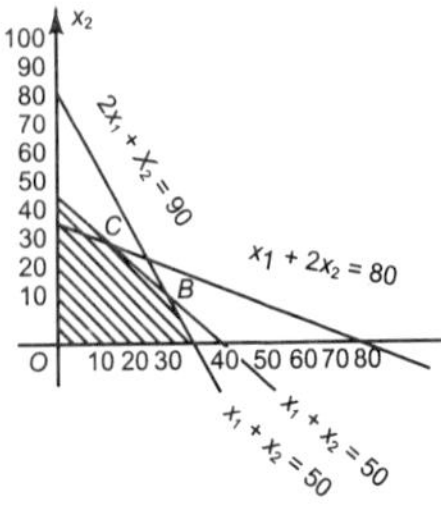

Fig. 33

The feasible region is shown by shaded portion.

The extreme points O, A, B, C, D are

$O = (0, 0), A = (45, 0)$
$B = (40, 10)$
$C = (20, 30)$ and $D = (0, 40)$

Let $Z = f(x_1, x_2) = 30x_1 + 40x_2$

$Z_O = 30 \times 0 + 40 \times 0 = 0$
$Z_A = f(45, 0) = 30 \times 45 + 40 \times 0 = 1350$
$Z_B = f(45, 10) = 30 \times 40 + 40 \times 10$
$= 1600$
$Z_C = f(20, 30) = 30 \times 20 + 40 \times 30$
$= 1800$
$Z_D = f(0, 40) = 30 \times 0 + 40 \times 40 = 1600$
Z max. $= 1800$ for $x_1 = 20$ $x_2 = 30$

37. Suppose that 8, 12 and 19 units of proteins carbohydrates and fats respectively are the minimum weekly requirements for a person Food *A* contains 2, 6 and 1 units of proteins carbohydrates and fats respectively per kilogram and food *B* contains 1, 1 and 3 units respectively per kilogram. If *A* costs Rs 0.85 per kg and *B* costs Rs. 0.40 per kg, how many kilograms of each should be buy per week to minimize his cost and still meet his minimum requirements.

Let the person purchase x_1 kg of food A and x_2 kg of food B. We construct the following table.

Ingradients	*Food*		*(Requirement)*
	A(x_1)	*B*(x_2)	
Proteins	2	1	8
Carbohydrates	6	1	12
Fats	1	3	9
Cost	Rs. 0.85	Rs. 0.40	

The mathematical formulation of the above L.P.P. is following

Min. $Z = 85x_1 + 40x_2$

s.t. $2x_1 + x_2 \geq 8$
$6x_1 + x_2 \geq 12$
$x_1 + 3x_2 \geq 9$
$x_1, x_2 \geq 0$

The shaded portion in the graphical representation shows the feasible region

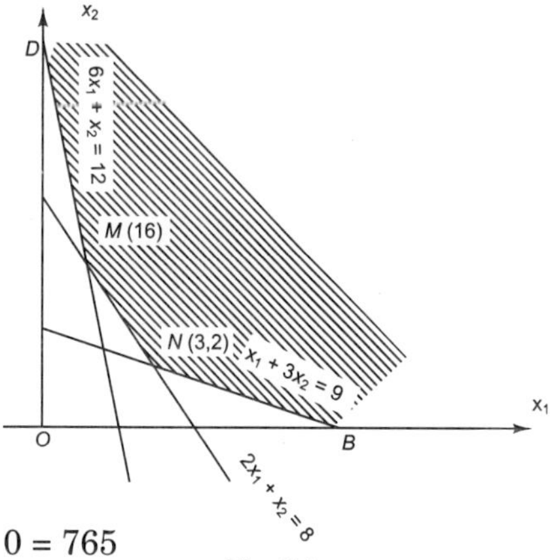

Fig. 34.

Let $Z = f(x_1, x_2) = 85x_1 + 40x_2$

$Z_B = f(9, 0) = 85 \times 9 + 40 \times 0 = 765$
$Z_N = f(3, 2) = 85 \times 3 + 40 \times 2 = 335$

$Z_M = f(1, 6) = 85 \times 1 + 40 \times 6 = 325$

$Z_D = f(0, 12) = 85 \times 0 + 40 \times 12 = 480$

Z min. $= 325$ for $x_1 = 1, x_2 = 6$

38. A man requires a certain minimum amount of each of 3 types of vitamins per day to maintain a normal health, let these 3 types of vitamins be found in specific quantities in our food items. The price per unit of each food item is known, find the combination of food items which supply the daily vitamin requirement and simultaneous by makes the cost of purchase minimum. The information needed are given in the following table.

Vitamin	*Vitamin content of food items per kg and price*		*Minimum daily requirements*
	Item I	*Item II*	
A_1	15 units	15 units	45 units
A_2	10 units	5 units	20 units
A_3	4 units	10 units	20 units
Price per kg	Rs. 5.0	Rs. 6.0	

Let x_1 kg of food item I and x_2 kg of food item II be required to met the requirement. There for the mathematical formulation is as below

$$\text{Min. } Z = 5x_1 + 6x_2$$
$$\text{s.t. } 15x_1 + 15x_2 \geq 45$$
$$10x_1 + 5x_2 \geq 20$$
$$4x_1 + 10x_2 \geq 20$$
$$x_1, x_2 \leq 0$$

The feasible region is the shaded portion

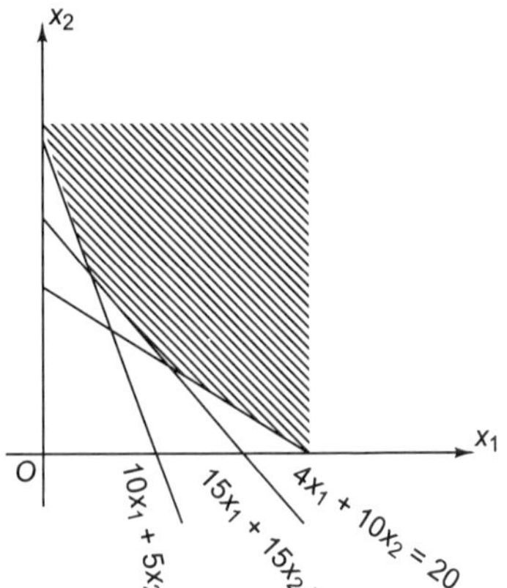

Fig. 35.

$$A = (5, 0)$$
$$B = \left(\frac{5}{3}, \frac{4}{3}\right)$$
$$C = (1, 2)$$
$$D = (0, 4)$$

Let $Z = f(x_1, x_2) = 5x_1 + 6x_2$

$$Z_A = f(5, 0) = 5 \times 5 + 6 \times 0 = 25$$
$$Z_B = f\left(\frac{5}{3}, \frac{4}{3}\right) = 5 \times \frac{5}{3} + 6 \times \frac{4}{3} = \frac{49}{3}$$
$$Z_C = f(1, 2) = 5 \times 1 + 6 \times 2 = 17$$
$$Z_D = f(0, 4) = 5 \times 0 + 6 \times 4 = 24$$
$$Z \text{ min.} = \frac{49}{3} \text{ for } x_1 = \frac{5}{3}, x_2 = \frac{4}{3}$$

39. A firm manufacturers two products *A* and *B* on which the profits earned per unit are Rs. 3 and Rs. 4 respectively. Each product is processed on two machines M_1 and M_2. Product *A* requires one minute of processing time on M_1 and two minutes on M_2 which *B* requires one minute on M_1 and one minute on M_2 ... M_1 is available for not more than 7 hours 30 minutes while machine, M_2 is available for 10 hours during any working day; find the number of units of product *A* and *B* to be manufactured to get the maximum profit.

Let x_1 and x_2 be the number of units of products *A* and *B* respectively to be manufactured.

Then the mathematical formulation is

$$\text{Max. } Z = 3x_1 + 4x_2$$

subject to
$$1x_1 + 1x_2 \le 450$$
$$2x_1 + 1x_2 \le 600$$
$$x_1, x_2 \ge 0$$

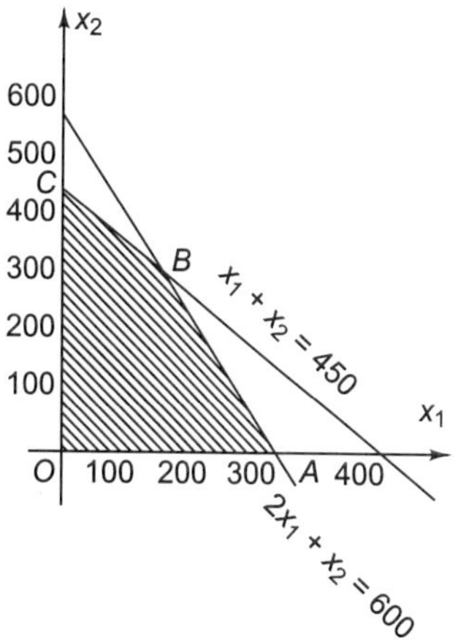

Fig. 36

The feasible region is shown by shaded portion

A is (300, 0)

B is (150, 300)

C is (0, 450)

Let $Z = f(x_1, x_2) = 3x_1 + 4x_2$

$Z_A = f(300, 0) = 3 \times 300 + 4 \times 0 = 900$

$Z_B = f(150, 300) = 3 \times 150 + 4 \times 300 = 1650$

$Z_C = f(0, 450) = 3 \times 0 + 4 \times 450 = 1800$

Z max. $= 1800$ for $x_1 = 0, x_2 = 450$.

40. Old hens can be bought for Rs. 2.00 each but young ones cost Rs. 5.00 each. The old hens lay 3 eggs per week and the young ones 5 eggs per week, each egg being worth 30 paise. A hen costs Rs. 1.00 per week to feed. If I have only Rs. 80.00 to spend for hens, how many of each kind should I buy to give a profit of more than Rs. 6.00 per week, assuming that I can not house more than 20 hens.

[Meerut L.P. 69, Bhopal (stat) 86, R.U.M.Sc. 82]

Solution: If x_1 as x_2 are the required numbers of old and young hens, the L.P.P. is

$$\text{Max. } Z = -0.1x_1 + 0.5x_2$$

s.t.
$$2x_1 + 5x_2 \le 80$$

$$x_1 + x_2 \le 20$$
$$x_1, x_2 \ge 0$$

Solution space is shown by shaded portion Extreme points are

$O = (0, 0)$

$A = (20, 0)$

$B = \left(\frac{20}{3}, \frac{40}{3}\right)$, $C = (0, 16)$

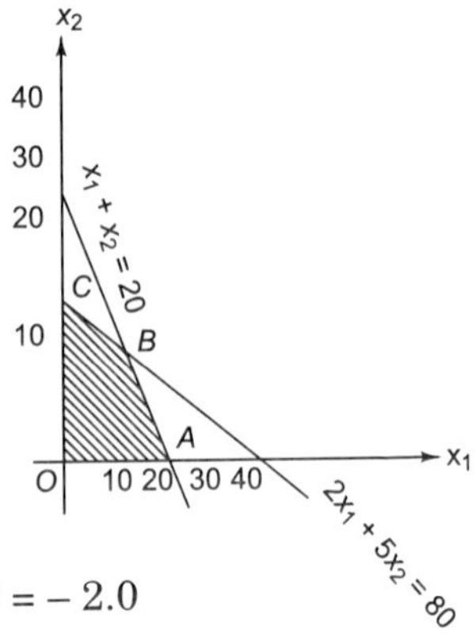

Fig. 37

Let $Z = f(x_1, x_2) = -0.1x_1 + 0.5x_2$

$$Z_A = f(20, 0) = -0.1 \times 20 + .5 \times 0 = -2.0$$

$$Z_B = f\left(\frac{20}{3}, \frac{40}{3}\right) = -.1 \times \frac{20}{3} + .5 \times \frac{40}{3}$$

$$= \frac{-2}{3} + \frac{20}{3} = 6$$

$$Z_C = f(0, 16) = -.1 \times 0 + .5 \times 16 = 8$$

Z max. $= 8$ for $x_1 = 0, x_2 = 16$

EXERCISE

1. Solve the following Linear programming Problems by graphical method:

(i) Max. $Z = 3x_1 + 2x_2$

subject to $2x_1 - x_2 \ge 2$

$x_1 + 2x_2 \le 8$

and $x_1, x_2 \ge 0$ [Delhi '73]

(ii) Max. $Z = 2x_1 + 3x_2$

subject to $x_1 + x_2 \le 1$

$3x_1 + x_2 \le 4$

and $x_1, x_2 \ge 0$ [Meerut '70]

(iii) Max. $Z = 5x_1 + 3x_2$

subject to $3x_1 + 5x_2 \le 15$

$5x_1 + 2x_2 \le 10$ [R.U.B.Sc. 86, 87]

and $x_1, x_2 \ge 0$ [Delhi M. Sc. (Math) '69, Dibrugrah M.Sc. (Stat. '76]

(iv) Max. $Z = 3x_1 + 2x_2$

subject to $-2x_1 + x_2 \le 1$

$x_1 \le 2$, $x_1 + x_2 \ge 3$

and $x_1, x_2 \ge 0$

(v) $\text{Max. } Z = 3x_1 + 4x_2$

subject to $x_1 - x_2 \leq -1, \quad -x_1 + x_2 \leq 0$

and $x_1, x_2 \geq 0$ [Meerut '74]

(vi) $\text{Max. } Z = 3x_1 + 2x_2$

subject to $-2x_1 - 3x_2 \leq 9, x_1 - 5x_2 \geq -20$

and $x_1, x_2 \geq 0$ [Madras B.E. '77]

(vii) $\text{Max. } Z = x_1 + x_2$

subject to $x_1 = 2x_2 \leq 2000$

$x_1 + x_2 \leq 1500$

$x \geq 600, x_1, x_2 \geq 0$

(viii) $\text{Max. } Z = 8000x_1 + 7000x_2$

subject to $3x_1 + x_2 \leq 66, \quad x_1 + x_2 \leq 45$

$x_1 \leq 20, x_2 \leq 40, x_1, x_2 \geq 0$ [Meerut (M.Sc.) '73]

(ix) $\text{Max. } Z = 3x_1 + 5x_2$

subject to $x_1 + 2x_2 \leq 2000, x_1 + x_2 \leq 1500$

$x_2 \leq 600$ and non-negativity restriction

2. Solve graphically the following L.P. Problems:

(i) $\text{Minimum } Z = -x_1 + 2x_2$

subject to $-x_1 + 3x_2 \leq 10$

$x_1 + x_2 \leq 6$

$x_1 - x_2 \leq 2$

and $x_1, x_2 \geq 0$

[Roorkee M.Sc. Math. 73]

(ii) $\text{Minimum } Z = 20x_1 + 10x_2$

subject to $x_1 - 2x_2 \leq 40$

$3x_1 + x_2 \geq 30$

$4x_1 + 3x_2 \geq 60$

and $x_1, x_2 \geq 0$ [Meerut M. Com. 77]

(iii) $\text{Minimum } Z = 2x_1 + 3x_2$

subject to $-x_1 + 2x_2 \leq 4,$

$x_1 + x_2 \leq 6$

$$x_1 + 3x_2 \geq 9$$

and $x_1, x_2 \geq 0$ [Delhi. '73]

(iv) Minimum $Z = 2x_1 + 3x_2$

subject to
$$x_1 + x_2 \leq 4$$
$$6x_1 + 2x_2 \geq 8$$
$$x_1 + 5x_2 \geq 4$$
$$x_1 \leq 3, x_2 \leq 3$$

and $x_1, x_2 \geq 0$

(v) Minimum $Z = 3x_1 + 5x_2$

subject to
$$-3x_1 + 4x_2 \leq 12$$
$$2x_1 - x_2 \geq -2$$
$$2x_1 + 3x_2 \geq 12$$
$$x_1 \leq 4, x_2 \geq 2$$

and $x_1, x_2 \geq 0$

(vi) Minimum $Z = 3x + 5y$

subject to
$$x + 3y \geq 3$$
$$x + y \geq 2$$

and $x, y \geq 0$

3. Solve the following L.P.P. graphically:

(i) Maximum $Z = 5x_1 + 4x_2$

subject to
$$x_1 - 2x_2 \leq 1$$
$$x_1 + 2x_2 \geq 3$$
$$x_1, x_2 \geq 0$$

(ii) Maximum $Z = 3x + 2y$

subject to
$$-2x + 3y \leq 9$$
$$3x - 2y \geq -20$$

and $x, y \geq 0$

(iii) Maximum $Z = 8x_1 + 7x_2$

subject to
$$3x_1 + x_2 \leq 66000$$
$$x_1 + x_2 \leq 45000$$
$$x_1 \leq 20000$$
$$x_2 \leq 40000$$

and $x_1 \geq 0, x_2 \geq 0$ [Meerut O.R. '73]

3. (a) What is meant by linear programming problem? Give a brief description of the problem with illustrations. How the same can be solved graphically? [Mysore B.Sc. (Stat) 1977]

4. A farmer has a 100 acre farm. He can sell the tomatoes, lettuce and radishes he can raise. The price he can obtain is Re. 1 per kg for tomatoes, Rs. 0.75 a head of lettuce and Rs. 2

per kg for radishes. The average yield per acre is 2000 kg of tomatoes, 3000 heads of lettuce and 1000 kg of radishes. Fertilizer is available as Rs. 0.50 per kg and the amount required per acre is 100 kg each for tomatoes and lettuce and 50 kg for radishes. Labour required for showing, cultivating and harvesting per acre 5 men/days for tomatoes and radishes and 6 men/days for lettuce. A total of 400 men/days of labour are available at Rs. 20/- per man/day.

Formulate this problem as a L.P.P. to maximize the farmer's total profit. [Delhi M.B.A. '76]

5. A company sells two different products A and B. The company makes a profit of Rs. 40 and Rs. 30 per unit on products A and B respectively. The two products are produced in a common production process and are sold in two different markets. It takes 3 hours to produce one unit of A and one hour to produce one unit of B. The production process has a capacity of 30000 man-hours. The market has been surveyed, and company officials feel that the maximum number of units of A that can be sold is 8,000 and the maximum of B is 12,000 units subject to these limitations, the products can be sold in any convex combination. Formulate the above L.P.P. and solve it by graphical method. [Delhi M.B.A. '77]
6. A factory uses three different resources for the manufacture of two different products, 20 units of the resource A, 12 units of the B and 16 units of C being available. One unit of the first product requires 2, 2 and 4 units of the respective resources and 1 unit of the second product requires 4, 2 and 0 units of the respective resources. It is known that the first produce gives a profit of 2 monetary unit per unit and the second 3. Formulate the L.P.P. How many units of each product should be manufactured for maximizing the profit? Solve it graphically.
7. A firm manufactures two types of products A and B and sells them at a profit of Rs. 2.00 on type A and Rs. 2.00 on type B. Each product as processed on two machines M_1 and M_2. Type A requires one minute of processing time on M_1 and two minutes on M_2, type B requires one minute on M_1 and one minute on M_2. The machine M_1 is available for more than 6 hours 40 minutes while machine M_2 is available for 10 hours during any working day.

   Formulate the problem as a L.P.P. and find how many products of each type should the firm produce each day in order to get maximum profit. [Dibrugran M.Sc. Math '75]
8. $$\text{Max. } Z = 10x_1 + 15x_2 + 20x_3$$

   s.t.
   $$10.7x_1 + 5x_2 + 2x_3 \le 2705$$
   $$5.4x_1 + 10x_2 + 4x_3 \le 2210$$
   $$0.7x_1 + x_2 + 2x_3 \le 445$$

ANSWERS

1. (i) $x_1 = 8, x_2 = 0, Z = 24$
   (ii) $x_1 = 0, x_2 = 1, Z = 3$
   (iii) $x_1 = 1.053, x_2 = 2.368, Z = 12.37$
   (iv) $x_1 = 2, x_2 = 1, Z = 8$
   (v) No solution
   (vi) No solution
   (vii) $x_1 = 1000, x_2 = 500, Z = 1500$
   (viii) $x_1 = 10.5, x_2 = 34.5, Z = 3255$
   (ix) $x_1 = 1000, x_2 = 500, Z = 5500$
2. (i) $x_1 = 2, x_2 = 0, Z = -2$
   (ii) $x_1 = 6, 2 = 12, Z = 240$
   (iii) $x_1 = 1.2, x_2 = 2.6, Z = 10.2$
   (iv) $x_1 = \frac{8}{7}, x_2 = \frac{4}{7}, Z = 4$
   (v) $x_1 = 3, x_2 = 2, Z = 19$
   (vi) Min. $Z = 7$, for $x = \frac{3}{2}, y = \frac{1}{2}$
3. (i) Unbounded solution
   (ii) No solution
   (iii) Max. $Z = 325500$ for $x_1 = 10500, x_1 = 34500$
   (iv) $x = 18, y = 12, Z = 72$
4. If the farmer sows tomatoes, lettue and radishes on x_1, x_2, x_3 acres respectively, then L.P.P.

   $$\text{Max. } Z = 1850x_1 + 2080x_2 + 1875x_3$$

   s.t. $x_1 + x_2 + x_3 \le 100,\ 5x_1 + 6x_2 + 5x_3 \le 400,$

   and $x_1, x_2, x_3 \ge 0$
5. If x_1 and x_2 are the units of products A and B respectively, L.P.P. is

   Max. $Z = 40x_1 + 30x_2$

   s.t. $3x_1 + x_2 \le 30000$

   $x_1 \le 8000, x_2 \le 12000$

   and $x_1, x_2 \ge 0$

Solution: $x_1 = 6000, x_2 = 12000$

Max. $Z = 400000$

6. First product = 2 units;
   Second product = 4 units
   Profit = 16 monentary units
7. L.P.P. is Max. $Z = 2A + 2B$

   s.t. $A + B \le 400$

   $2A + B \le 600$

   $A, B \ge 0$

   $A = 200, B = 200, Z = 1800$

   or $A = 0, B = 400, Z = 800$
8. $x_1 = 200, x_2 = 65, x_3 = 120$

   $Z = 5375$

2

Simplex Method

2.1. INTRODUCTION

Simplex method (also called Simplex technique or simplex algorithm) was developed by G.B. Dantzig. It has the advantage of being universal that, if any linear model for which the solution exists, can be solved by it. In principle it consists of starting with a certain solution of which all that we know is that it is feasible (that is, it satisfies the non-negativity conditions $\chi_i \geq 0$ ($i = 1, 2, 3, \ldots n$). We then improve upon this solutions at consecutive stages, until a certain finite number of stages we arrive at the optimal solutions. The basis of the simplex method consists of two fundamental conditions.

(i) The feasibility conditions—It ensures that if the starting solutions is basic feasible, only basic solutions will be obtained during computations.

(ii) The optimality conditions—It guarantees that only better solutions (as compared to the current solutions) will be encountered.

The simplex method is an iterative procedure which either involves a L.P.P. in a finite number of steps or gives an indication that there is an unbounded solution to the L.P.P. Firstly, some terms and definitions will be explained and then the computational techniques will be explained with the help of suitable examples.

2.2. SLACK AND SURPLUS VARIABLES

[R.U.B. SC. '88]

If a constraint has a sign ≤, then in order to make it an equality we have to add some thing to the left hand side.

The variables which are added to L.H.S. to the constraints to convert them into equalities are called the slack variables.

For example, consider the L.P.P.

Max. $z = x_1 + 3x_2 + 2x_3$

$$\left.\begin{aligned} x_1 + x_2 + x_3 &\le 3 \\ 2x_1 + 5x_2 - x_3 &\le 5 \end{aligned}\right\} \quad \dots (1)$$

$$x_1, x_2, x_3 \ge 0$$

In order to convert constraints (1) into equalities (equations) we add x_4 and x_5 on L.H.S. of (1). Then we have

$$x_1 + x_2 + x_3 + x_4 = 3$$

$$2x_1 + 5x_2 - x_3 + x_5 = 6$$

Here, x_4 and x_5 are called slack variables.

(i) Surplus Variable— If a constraint has a sign ≥, then in order to make it an equality we have to subtract some thing from its L.H.S.

The variables which are subtracted from L.H.S. of the constraints to convert them into equalities are called the surplus variables.

For example—

$$\left.\begin{aligned} &\text{Mini.} && Z = 3x_1 + 4x_2 + 5x_3 \\ &\text{Subject to} && x_1 + 2x_2 + 3x_3 \ge 6 \\ &&& 2x_1 + 3x_2 + x_3 \ge 2 \\ &&& 2x_1 - 2x_2 + 4x_3 \ge 3 \end{aligned}\right\} \quad \dots (2)$$

Inequalities (2) can be changed to equalities by subtracting non-negative variables x_4, x_5, x_6 from L.H.S. Thus we have

$$x_1 + 2x_2 + 3x_3 - x_4 = 6$$

$$2x_1 + 3x_2 + x_3 - x_5 = 2$$

$$2x_1 - 2x_3 + 4x_3 - x_6 = 3$$

x_4, x_5, x_6 are called surplus variables.

Note→ (1) Slack = Requirement – Production

(2) Surplus = Production – Requirement.

2.3. THE GENERAL MATHEMATICAL FORMULATION FOR LINEAR PROGRAMMING

Maximum or minimize the objective function

$$z = c_1x_1 + c_2x_2 + \ldots + c_nx_n \quad \ldots(1)$$

Subject to

$$\left.\begin{array}{l} a_{11}x_2 + a_{12}x_2 + \ldots + a_{1n}x_n \ (\leq, =, \geq)\ b_1 \\ a_{21}x_1 + a_{22}x_2 + \ldots + a_{2n}x_n \ (\leq, =, \geq)\ b_2 \\ \ldots \quad \ldots \quad \ldots \quad \ldots \quad \ldots \quad \ldots\ldots\ldots\ldots \\ a_{m1}x_1 + a_{m2}x_2 + \ldots + a_{mn}\, x_n \ (\leq, =, \geq) b_m \end{array}\right\} \quad \ldots (2)$$

and $\quad x_1, x_2 \ldots x_n \geq 0 \ldots \ldots \ldots \quad \ldots (3)$

The *Objective function z* may be profit, loss, cost etc.

$x_1, x_2, \ldots x_n$ are called decision variables.

$c_1, c_2, \ldots c_n$ are called cost co-efficient $b_1, b_2 \ldots b_m$ are called Stipulations.

$a_{11}, a_{12}, a_{21}, a_{22} \ldots$ etc. that is co-efficients of $x_1, x_2, \ldots x_n$ in constraints are called *structural* co-efficients.

Some Definitions—In (2) out of ($\leq, \geq, =$) one sign is taken.

Solution: $x_i (i = 1, 2, \ldots n)$ is a solution of the above general L.P.P. if it satisfies (2) and (3).

Non-degenerate Solution: A basic solution is called non-degenerate basic solution of none of the basic variables is zero.

Degenerate Solution: A basic solution is called Degenerate if one or more of the two basic variables vanish.

Feasible Solution: A feasible solution to a L.P.P. is the set of values of the variables which satisfies the constraints and non-negative conditions of the problem.

Basic feasible Solution: A feasible solution which is also basic is called a basic feasible solution. [R.U.B. Sc.87]

Non-degenerate basic feasible solution: A basic feasible solution is said to be non-degenerate B.F.S. if none of the basic variables vanishes.

Optimal Solution: A basic feasible solution is said to be optimal or optimum if it also optimizes the objective function while satisfying the constraints and non-negativity of decision variables.

These definitions can be well understood with the following example.

Sample Problems

Example 1. Find all the basic solutions to the following problem.

$$x_1 + 2x_2 + 3x_3 = 4$$

$$2x_1 + 3x_2 + 5x_3 = 7$$

Also, find which of the basic solutions are

(i) basic feasible

(ii) non-degenerate-basic feasible

[V.B.U. Part III Hons '99]

Solution: Here, n (number of decision variables) = 3, m (number of constraints) = 2. A basic solution can be found by setting any one of the ($n - m = 3 - 2 = 1$) variables equal to zero, and then solving the constraints equations. The total number of basic solutions is

$$^nC_m = {}^3C_2 = \frac{3!}{2!\,1!} = 3$$

Putting $x_2 = 0$, in the constraints and

Solving these two equations, $x_1 = 2, x_2 = 1$

Thus a basic (non-basic) solution to the given system is

(i) (Basic) : $x_1 = 2, x_2 = 1$. Non basic $x_3 = 0$

Similarly, the other two basic and non-basic solutions are

(ii) (Basic) : $x_1 = 1, x_3 = 1$: Non-basic $x_2 = 0$

(iii) (Basic) : $x_2 = -1; x_3 = 2$; Non-basic $x_1 = 0$.

The first basic solution is *basic feasible* as both the basic variable $x_1, x_2 \geq 0$.

The second basic solution is also basic feasible as $x_1, x_3 \geq 0$.

But the third basic solution is not Basic Feasible Solution as $x_2 < 0$. Also, the first two basic feasible solutions are Non-degenerate basic feasible as they have exactly two basic variables x_1 and x_2 in (i) and x_1, x_3 in (ii) are > 0.

2. Obtain the basic solution in two following system of linear equations

$$x_1 + 2x_2 + x_3 = 4$$

$$2x_1 + x_2 + 5x_3 = 5$$

[Meerut M.Sc. Math 74, R.U. 82 M.Sc., R.U.B. Sc. 87, R.U. B.Sc. Hons. 87, V.B.U. Part II Hons 99]

Also, indicate which of these basic solutions are

(i) basic feasible

(ii) degenerate basic feasible

(iii) non-degenerate basic feasible.

Solution: Here, n = number of variables = 3

m = number of constraints = 2

By setting any of the ($n - m = 3 - 2 = 1$) variables equal to zero and then solving we get basic solution. The number of

Basic Solution = $^nC_m = 3C_2 = 3$

Putting $\quad x_3 = 0$ and solving $x_1 = 2, x_2 = 1$

$x_2 = 0$ and solving $x_1 = 5, x_3 = -1$

$x_1 = 0$ and solving $x_2 = 5/3, x_3 = 2/3$.

Thus the Solution are

(a) Basic $x_1 = 2, x_2 = 1,\quad$ non-basic $x_3 = 0$

(b) Basic $x_1 = 5, x_3 = -1,\quad$ non-basic $x_2 = 0$

(c) Basic $x_2 = 5/3, x_3 = 2/3,\quad$ non-basic $x_1 = 0$

(a) and (c) are non-degenerate basic feasible solutions. Clearly, the solution (5, 0, −1) is not a feasible solutions. Here there are only two basic feasible solutions.

$\left(0, \frac{5}{3}, \frac{2}{3}\right)$ and (2, 1, 0). Hence there are two non-degenerate basic feasible solutions; but three non-degenerate basic solution.

3. For the L.P.P.

Maximize $\quad z = 12x_1 + 8x_2 + 14x_3 + 10x_4$

Subject to $\quad 5x_1 + 4x_2 + 2x_3 + x_4 = 100$

$2x_1 + 3x_2 + 8x_3 + x_4 = 75$

$x_1, x_2, x_3\ x_4 \geq 0$

Find

(i) basic solution

(ii) basic feasible solution

(iii) non-degenerate basic feasible solution

(iv) optimal basic feasible solution.

Solution: Here, $n = 4$, $m = 2$. A basic solution can be obtained by setting any of the ($n - m = 2$) non-basic variables equal to zero and then solving the constraint (containing the basic variables) equations. The total number of basic solution is

$$^4C_2 = \frac{4!}{2!\,2!} = 6$$

(i) Put $x_3 = 0$. $x_4 = 0$. The equations become

$$5x_1 + 4x_2 = 100$$

$$2x_1 + 3x_2 = 75$$

$\therefore \quad x_1 = 0, x_2 = 25$

Basic solution $x_1 = 0, x_2 = 25$; Non-basic $x_3 = 0, x_4 = 0$

$\therefore \quad z = 12x_1 + 8x_2 + 14x_3 + 10x_4$; $z = 200$

(ii) Put $x_2 = 0, x_4 = 0$. Then, $5x_1 + 2x_3 = 100$

$2x_1 + 8x_3 = 75$

$\therefore \quad x_1 = \frac{325}{18} \quad x_3 = \frac{175}{36} \quad \therefore \quad z = \frac{5175}{18}$

Basic $\quad x_1 = \frac{325}{18} \quad x_3 = \frac{175}{36}$

Non-basic $\quad x_2 = 0, \quad x_4 = 0, \quad z = \frac{5125}{18}$

(iii) Put $x_2 = 0, x_3 = 0$ in the given equations and

Solving, $\quad x_1 = \frac{25}{3} \quad x_4 = \frac{175}{3}$

Basic $\quad x_1 = \frac{25}{3}, \quad x_4 = \frac{175}{3}$

Non-basic $\quad x_2 = 0, x_3 = 0$

$$z = \frac{2050}{3}$$

(iv) $x_1 = 0, x_4 = 0$; Then $x_2 = 25, x_3 = 0$

Basic $\quad x_2 = 25, x_3 = 0$

Non-basic $\quad x_1 = 0, x_4 = 0$

$$z = 200.$$

(v) Put $x_1 = 0, x_3 = 0$ $\qquad$ Then $x_2 = 25, x_4 = 0$

Basic $\quad x_2 = 25, \qquad x_4 = 0$

Non-basic $\quad x_1 = 0, \qquad x_2 = 0$

$$z = 200$$

(vi) Put $x_1 = 0, x_2 = 0$ then $x_3 = -\frac{25}{6}$

$$x_4 = \frac{325}{3}$$

$$z = 1025$$

Basic $\quad x_3 = \frac{-25}{6} \quad x_4 = \frac{325}{3}$

Non-basic $\quad x_1 = 0, x_2 = 0.$

(i) to (vi) are Basic solutions. Basic feasible solutions are (i), (ii), (iii), (iv) and (v) as all basic solutions in these are either zero or positive.

Non-degenerate basic feasible solutions are (ii) and (iii) as they have exactly m (here $m = 2$) positive (non-zero) basic solution. Optimal basic feasible solution is (iii)

Which gives $\quad x_1 = \frac{25}{3}, x_2 = 0, x_3 = 0, x_4 = \frac{175}{3}$

$$Z_{\max} = \frac{2050}{3}$$

4. Find all the basic solutions possible of the system

$$2x_1 + 3x_2 + 5x_3 + 6x_4 = 16$$
$$x_1 + 2x_2 + 2x_3 + 3x_4 = 9 \qquad ...(1)$$

Since there are 2 equations in four unknown variables there will be ${}^4C_2$ (i.e., 6) different pairs of equation in two unknown possible. Below we shall write all these possible pairs and the basic solution of the system (1)

(i) $2x_1 + 3x_2 = 16$, $x_1 + 2x_2 = 9$ $\quad x_1 = 5, x_2 = 2, (x_3 = x_4 = 0)$ ...(2)

(ii) $2x_1 + 5x_3 = 16$, $x_1 + 2x_3 = 9$ $\quad x_1 = 13, x_3 = -9\ (x_2 = x_4 = 0)$ ...(3)

(iii) $2x_1 + 6x_4 = 16$, $x_1 + 3x_4 = 9$ $\quad$ No solution exists as the two lines are parallel, also note the equations are inconsistent. ...(4)

(iv) $3x_2 + 5x_3 = 16$, $2x_2 + 2x_3 = 9$ $\quad x_2 = \frac{13}{4}\quad x_3 = \frac{5}{4}\ (x_1 = x_4 = 0)$ ...(5)

(v) $3x_2 + 6x_4 = 16$, $2x_2 + 3x_4 = 9$ $\quad x_2 = 2, x_4 = \frac{5}{4}\ (x_1 = x_3 = 0)$ ...(6)

(vi) $5x_3 + 6x_4 = 16$, $2x_3 + 3x_4 = 9$ $\quad x_3 = -2, x_4 = \frac{13}{3}\ (x_1 + x_2 = 0)$ ...(7)

Thus, there are 5 solutions of above problem. We should not have considered the equation (4) because

$$A = \begin{bmatrix} 2 & 6 \\ 1 & 3 \end{bmatrix} \text{ is singular.}$$

5. Show that two of the basic solutions of the system

$$3x_1 + 4x_3 + x_4 = 12$$
$$5x_1 + 4x_2 + 2x_3 = 20 \qquad ...(1)$$

are degenerate

Proceeding as before we get ${}^3C_2$ (i.e. 3) sets of equations

(i) $3x_1 + 0x_2 = 12$, $5x_1 + 4x_2 = 20$ $\quad$ i.e. $\quad x_1 = 4, x_2 = 0$ ...(2)

(ii) $3x_1 + 4x_3 = 12$, $5x_1 + 2x_3 = 20$ $\quad$ i.e. $\quad x_1 = 4, x_3 = 0$ ...(3)

(iii) $0x_2 + 4x_3 = 12$, $4x_2 + 2x_3 = 20$ $\quad$ i.e. $\quad x_2 = \frac{7}{2}, x_3 = 3$ ...(4)

Some Useful Notations

Here we shall introduce some basic notations which are frequently used in our discussion. Consider the following L.P.P.

Max. $\quad Z = c_1x_1 + c_1x_2 + \ldots + c_nx_n$

Subject to constraints

$$a_{11}x_1 + a_{12}x_2 + \ldots + a_{1n}x_n \le b_1$$
$$a_{21}x_1 + a_{22}x_2 + \ldots + a_{2n}x_n \le b_2$$
$$\ldots \quad \ldots \quad \ldots \quad \ldots$$
$$a_{m1}x_1 + a_{m2}x_2 + \ldots + a_{mn}x_n \le b_m$$
$$x_1, x_2 \ldots, x_n \ge 0.$$

After introducing the slack variables $x_{n+1}, x_{n+2}, \ldots x_{n+m}$, the above L.P.P. takes the form

$$\text{Max.} \quad Z = c_1x_1 + c_2x_2 + \ldots c_nx_n + 0x_{n+1} + 0x_{n+2} + \ldots + 0x_{n+m}$$
$$a_{11}x_1 + a_{22}x_2 + \ldots + a_{1n}x_n + x_{n+1} = b_1$$
$$\ldots \quad \ldots \quad \ldots \quad \ldots$$
$$a_{21}x_1 + a_{22}x_2 + \ldots + a_{2n}x_n + x_{n+2} = b_2$$
$$a_{m1}x_1 + a_{m2}x_2 + \ldots a_{mn}x_n + x_{n+m} = b_m$$
$$x_1, x_2, \ldots x_n, x_{n+1} \ldots x_{n+m} \ge 0$$

In the matrix notation, the above L.P.P. is as follows

Max. $Z = cx$

Subject to $Ax = b,$

$x = 0$

$$\text{where } A = \begin{bmatrix} a_{11} & a_{12} \ldots a_{1n} & 1 & 0 & 0 & \ldots\ldots & 0 \\ a_{21} & a_{22} \ldots a_{2n} & 0 & 1 & 0 & \ldots\ldots & 0 \\ \ldots & \ldots\ldots\ldots\ldots & \ldots & \ldots & \ldots & \ldots & \ldots \\ a_{m1} & a_{m2} & a_{mn} & 0 & 0 & \ldots & 1 \end{bmatrix}_{m \times (m+n)}$$

$$x = \begin{bmatrix} x_1 \\ x_2 \\ \ldots \\ x_n \\ x_{n+1} \\ \ldots\ldots \\ X_{n+m} \end{bmatrix}_{(n+m) \times 1} = (x_1, x_2, x_n, x_{n+1}, \ldots, x_{n+m})^T$$

$$b = \begin{bmatrix} b_1 \\ b_2 \\ \ldots \\ b_m \end{bmatrix}_{m \times 1} = (b_1, b_2, \ldots b_m)^{\text{T}}.$$

$$C = (c_1, c_2, \ldots, c_n, 0, 0 \ldots 0)_{1 \times (m+n)}$$

Let a_1 denote the ith column of the matrix A.

$$a_i = \begin{bmatrix} a_{1i} \\ a_{2i} \\ \ldots \\ b_{mi} \end{bmatrix}$$

$$a_1 = \begin{bmatrix} a_{11} \\ a_{21} \\ \cdots \\ a_{m1} \end{bmatrix}, \quad a_2 = \begin{bmatrix} a_{12} \\ a_{23} \\ \cdots \\ a_{m2} \end{bmatrix}, \quad a_3 = \begin{bmatrix} a_{23} \\ a_{22} \\ \cdots \\ a_{m3} \end{bmatrix}$$

$$a_n = \begin{bmatrix} a_{1n} \\ a_{2n} \\ \cdots \\ a_{mn} \end{bmatrix}, a_{n+1} = \begin{bmatrix} 1 \\ 0 \\ \cdots \\ 0 \end{bmatrix}, \ldots, a_{n+m} = \begin{bmatrix} 0 \\ 0 \\ \cdots \\ 1 \end{bmatrix}$$

Thus A can be written is $A = [a_1, a_2, \ldots a_n, a_{n+1}, \ldots, a_{n+m}]$

$$\text{Clearly, } [a_{n+1}, a_{n+2}, \ldots a_{n+m}] = \begin{bmatrix} 0 & 1 & \ldots & 0 \\ 0 & 1 & \ldots & 0 \\ \ldots & \ldots & \ldots & \ldots \\ 0 & 0 & & 1 \end{bmatrix}$$

I. Basic Matrix

We denote B by $m \times m$ non-singular matrix chosen from A where column vectors are m number of linearly independent (L.I.) column of A.

i.e. determinant $B \neq 0$

i.e. $| B | \neq 0$

i.e. B^{-1} exists.

Then B is called the Basic matrix.

But we know that the maximum possible number of submatrix B is

$$^{n+m}C_m = \frac{m+n!}{m!\,n!}$$

because out of $(n + m)$ column of A, m columns of B can be selected in $^{n+m}C_m$ ways.

Denote the columns of B by $\beta_1, \beta_2, \beta_3, \ldots \beta_m$. Then

$$B = (\beta_1, \beta_2, \beta_3, \ldots, \beta_m)$$

Obviously, the vectors $\beta_1, \beta_2, \beta_3, \ldots, \beta_m$ are remaining or rearrangement of the columns $(a_1, a_2, a_3, \ldots a_{n+m})$ of A. When the m column vectors are selected out of $(n + m)$ column vectors in A. Thus, we find that m members of A are selected from $(a_1, a_2, a_3, \ldots, a_{n+m})$ to from the basis B and are given a name $\beta_1, \beta_2, \beta_3 \ldots, \beta_m$.

II. Basic Variables

The variables corresponding to $\beta_1, \beta_2, \beta_3, \ldots \beta_m$ are called Basic Variables and are denoted by $x_{B1}, x_{B2}, x_{B3}, \ldots, x_{Bm}$ respectively. The vector of these m basic variables is denoted by X_B so that

$$X_B = (x_{B_1}, x_{B_2}, x_{B_3}, ..., x_{B_m})$$

Thus x_{Bi} is a basic variable where as X_B is a vector of basic variables.

The subscript x_{Bi} indicates that x_{Bi} correspond to column β_i of basic matrix B.

Let us take up the following example to understand these notations.

Consider the following system of linear equations

$$x_1 + 2x_2 + x_3 = 4$$
$$2x_1 + x_2 + 5x_3 = 5$$

These two equations can be put in the matrix from as $Ax = b$ where

$$A = \begin{bmatrix} 1 & 2 & 1 \\ 2 & 1 & 5 \end{bmatrix}, \quad x = \begin{bmatrix} x_1 \\ x_2 \\ x_3 \end{bmatrix} \quad b = \begin{bmatrix} 4 \\ 5 \end{bmatrix}$$

A can be written as

$A = (a_1, a_2, a_3)$ where

$$a_1 = \begin{bmatrix} 1 \\ 2 \end{bmatrix}, \quad a_2 = \begin{bmatrix} 2 \\ 1 \end{bmatrix}, \quad a_3 = \begin{bmatrix} 1 \\ 5 \end{bmatrix}$$

In this problem there are three Basic matrices, namely,

$$B_1 = \begin{bmatrix} 1 & 2 \\ 2 & 1 \end{bmatrix} = (a_1, a_2)$$

$$B_2 = \begin{bmatrix} 1 & 1 \\ 2 & 5 \end{bmatrix} = (a_1, a_3)$$

$$B_3 = \begin{bmatrix} 2 & 1 \\ 1 & 5 \end{bmatrix} = (a_2, a_3)$$

Out of these three i.e. (B_1, B_2, B_3) any one of these can be taken as B.

Let $B = B_2$, Then

$$B = [\beta_1, \beta_2], \text{ where } \beta_1 = \begin{bmatrix} 1 \\ 2 \end{bmatrix}, \quad \beta_2 = \begin{bmatrix} 1 \\ 5 \end{bmatrix}$$

The basic variables associated with the submatrix B_1 are x_1, x_2, the basic variables associated with submatrix B_2 are x_1, x_2 and the basic variables associated with the submatrix B_3 are x_2, x_3.

On account of this, the equations giving the basic solutions can be written as

(i) $$\begin{bmatrix} 1 & 2 \\ 2 & 1 \end{bmatrix} \times \begin{bmatrix} x_1 \\ x_2 \end{bmatrix} = \begin{bmatrix} 4 \\ 5 \end{bmatrix}$$

(ii) $$\begin{bmatrix} 1 & 1 \\ 2 & 5 \end{bmatrix} \times \begin{bmatrix} x_1 \\ x_3 \end{bmatrix} = \begin{bmatrix} 4 \\ 5 \end{bmatrix}$$

(iii) $\begin{bmatrix} 2 & 1 \\ 1 & 5 \end{bmatrix} \times \begin{bmatrix} x_2 \\ x_3 \end{bmatrix} = \begin{bmatrix} 4 \\ 2 \end{bmatrix}$

Now, denote by X_{B1}, X_{B2}, X_{B3} the

Vectors of the basic variables associated to the submatrices B_1, B_2, B_3 respectively, that is,

$$^XB_1 = \begin{bmatrix} x_1 \\ x_2 \end{bmatrix}, \ ^XB_2 = \begin{bmatrix} x_1 \\ x_3 \end{bmatrix}$$

and $$^XB_3 = \begin{bmatrix} x_2 \\ x_3 \end{bmatrix}$$

Using these notations, (i), (ii), and (iii) can be written as

$$B_1 {}^XB_1 = b$$
$$B_2 {}^XB_2 = b$$
$$B_3 {}^XB_3 = b$$

But B_1, B_2, B_3 are non-singular

$\therefore$ $$^XB_1 = B_1^{-1} b$$
$$^XB_2 = B_2^{-1} b$$
$$^XB_3 = B_3^{-1} b$$

Let B the basic matrix and let us denote the remaining column vectors of A by S. Thus, we have $A = (B, S)$.

Let x be the basic solution of $Ax = b$ which is $(n + m)$ component vectors. Thus both zero (non-basic) and non-zero (non-basic) variables are contained in x. Now, let us suppose that the components of x are so arranged that the first m components are basic variables say

$x_{B1}, x_{B2}, x_{B3}, \ldots, x_{Bm}$ and hence denoted by X_B.

Therefore $x = (x_{B1}, x_{B2}, x_{B3}, \ldots x_{Bm}, 0, 0, \ldots 0)$

i.e. $= (X_B, 0)$ where

O is a n-Component null Vector.

So, $Ax = b$ becomes

$(B, S)\,(X_{Bo}) = b$

$\Rightarrow BX_B = b$

$\Rightarrow X_B = B^{-1} b$ as B is non-singular.

III. Notations CBr

The Co-efficients of basic variables $x_{B1}, x_{B2}, x_{B3}, \ldots, x_{Bm}$ in the Objective function

$$Z = c_1x_1 + c_2x_2 + c_3x_3 + \ldots + c_nx_n + 0x_{n+1} + \ldots + 0x_{n+m}$$

are denoted by $C_{B_1}, C_{B_2}, C_{B_3}, \ldots, C_{B_m}$ and written as

$$C_B = (C_{B_1}, C_{B_2}, C_{B_3}, \ldots C_{B_m})$$

Clearly, C_B is the m-Component row-vector containing the constants taken from the objective function associated with the basic variables C_{B_i} is the Co-efficient of the basic variable x_{Bi} in the objective function.

Under the notations

$$Z = C_{B1}\, x_{B1} + C_{B2}\, x_{B2} + \ldots + C_{Bm}\, x_{Bm} + 0$$

Since all non-basic variables are zero (because all remaining n variables are non-basic and hence zero)

$$Z = (C_{B1}, C_{B2}, \ldots C_{Bm})\,(x_{B1}, x_{B2}, \ldots x_{Bm})$$

i.e. $\qquad Z = C_B X_B$

IV. Symbol *Yi*

Since $B = (B_1, B_2, \ldots B_m)$ is a non-singular matrix $(|B| \neq 0)$ of order $m \times m$ and β_1 the vectors $\beta_1, \beta_2, \ldots \beta_m$ are linearly independent column chosen from matrix A, they from the basis of S^m. Hence each vector in S^m and so any column a_j of A can be expressed as a linear combinations of the vectors (columns) of B.

Let $\qquad a_i = y_{1i}\, \beta_1 + y_{2i}\, \beta_2 \ldots + y_{mi}\, \beta_m$

Where $\qquad y_{1i}\, y_{2i} \ldots, y_{mi}$ are scalars

$$= (\beta_1, \beta_2, \beta_3, \ldots, \beta_m) \begin{bmatrix} y_{1i} \\ y_{2i} \\ \ldots\ldots \\ y_{mi} \end{bmatrix}$$

or, $\qquad a_i = BY_i$ where $Y_i = (y_{1i}, y_{2i}, \ldots y_{mi})^T$

$\therefore \qquad Y_i = B^{-1}\, a_i$

The vector Y_i will change if the columns of A, making up B will change.

To have clear idea of these notations consider the following example:

Max. $\quad Z = 7x_1 + 5x_2$

Subject to $\qquad x_1 + 2x_2 \leq 6$

$\qquad 4x_1 + 3x_2 \leq 12$

$\qquad x_1, x_2 \geq 0$

The given problem by the use of slack variables x_3 and x_4 can be converted into equations as follows

Max. $\qquad Z = 7x_1 + 5x_2 + 0x_3 + 0x_4$

$$\left.\begin{aligned} x_1 + 2x_2 + x_3 &= 6 \\ 4x_1 + 3x_2 + x_4 &= 12 \end{aligned}\right\} \quad \ldots(1)$$

(1) can be put as

$$\left.\begin{aligned} x_1 + 2x_2 + x_3 + 0x_4 &= 6 \\ 4x_1 + 3x_2 + 0x_3 + x_4 &= 12 \end{aligned}\right\} \quad ...(2)$$

(2) can be written as

$$\begin{bmatrix} 1 & 2 & 1 & 0 \\ 4 & 3 & 0 & 1 \end{bmatrix} \begin{bmatrix} x_1 \\ x_2 \\ x_3 \\ x_4 \end{bmatrix} = \begin{bmatrix} 6 \\ 12 \end{bmatrix}$$

or, $Ax = b$, where $A = \begin{bmatrix} 1 & 2 & 1 & 0 \\ 4 & 3 & 0 & 1 \end{bmatrix}$, $x = \begin{bmatrix} x_1 \\ x_2 \\ x_3 \\ x_4 \end{bmatrix}$, $b = \begin{bmatrix} 6 \\ 12 \end{bmatrix}$

or, $[a_1, a_2, a_3, a_4]x = b$

where $a_1 = \begin{bmatrix} 1 \\ 4 \end{bmatrix}$, $a_2 = \begin{bmatrix} 2 \\ 3 \end{bmatrix}$, $a_3 = \begin{bmatrix} 1 \\ 0 \end{bmatrix}$, $a_4 = \begin{bmatrix} 0 \\ 1 \end{bmatrix}$

The number of basis matrices $= {}^4C_2 = \dfrac{4!}{2!\,2!} = 6$

Obviously the rank of matrix A is = 2

Since $a_3 = \begin{bmatrix} 1 \\ 0 \end{bmatrix}$ and $a_4 = \begin{bmatrix} 0 \\ 1 \end{bmatrix}$ are two linearly independent column vectors of A, we take

$$B = (\beta_1, \beta_2) = \begin{bmatrix} a_3 & a_4 \\ 1 & 0 \\ 0 & 1 \end{bmatrix}$$

as a non-singular submatrices of A,

B is non-singular as $|B| = \begin{vmatrix} 1 & 0 \\ 0 & 1 \end{vmatrix} = 1 \neq 0$

Clearly $B = I_2$ is a unit vector of order 2.

The basic variables associated in the β_1 and β_2 are those which are associated with a_3 and a_4 and hence they are x_3 and x_4 and

$$X_B = B^{-1} b$$

$$= \begin{bmatrix} 1 & 0 \\ 0 & 1 \end{bmatrix} \begin{bmatrix} 6 \\ 12 \end{bmatrix} = (6, 12)$$

$$\left(B^{-1} = \text{Inverse of } B = \begin{bmatrix} 1 & 0 \\ 0 & 1 \end{bmatrix} \text{ is } \begin{bmatrix} 1 & 0 \\ 0 & 1 \end{bmatrix} \right)$$

or, $(x_{B_1}, x_{B_2}) = (6, 12)$

$x_{B_1} (= x_3) = 6$

and $x_{B_2} (= x_4) = 12$

$$C_B = (C_{B_1}, C_{B_2})$$
$$= (\text{coeff. of } x_3 \text{ in objective function}$$
$$\text{coeff. of } x_4 \text{ in objective function})$$
$$= (0, 0)$$

Value of Z for $x_1 = 0, x_2 = 0, x_3 = 6, x_4 = 12$ is

$$Z = C_B X_B = (0, 0)\begin{bmatrix} 6 \\ 12 \end{bmatrix} = C$$

$$Y_1 = B^{-1} a_1 = \begin{bmatrix} 1 & 0 \\ 0 & 1 \end{bmatrix} \begin{bmatrix} 1 \\ 4 \end{bmatrix} = \begin{bmatrix} 1 \\ 4 \end{bmatrix} = a_1$$

$$Y_2 = B^{-1} a_2 = \begin{bmatrix} 1 & 0 \\ 0 & 1 \end{bmatrix} \begin{bmatrix} 2 \\ 3 \end{bmatrix} = \begin{bmatrix} 2 \\ 3 \end{bmatrix} = a_2$$

Some Useful Theorems

Theorem (*Extreme Point Correspondence*): A basic feasible solution to an L.P.P. must correspond to an extreme point of the set of all feasible solutions.(V.B.U. Part III Hons. '98)

Proof: Let the L.P.P. be determined x so as to maximize,

$$z = cx, x^T \in R^n$$

subject to the constraints:

$$Ax = b$$

and $x \geq 0$, where $A = \begin{bmatrix} a_{11} & a_{12} & \cdots\cdots & a_{1n} \\ a_{21} & a_{22} & \cdots\cdots & a_{2n} \\ \cdots\cdots & \cdots\cdots & \cdots\cdots & \cdots\cdots \\ a_{m1}, & a_{m2}, & & a_{mn} \end{bmatrix}, x = \begin{bmatrix} x_1 \\ x_2 \\ \cdots \\ x_n \end{bmatrix}, b = \begin{bmatrix} b_1 \\ b_2 \\ \cdots \\ bm \end{bmatrix}$

$$C = [c_1, c_2, \ldots c_m]$$

Let $\rho(A) = m$, where $\rho(A)$ means rank of A.

Let S be the set of all feasible solutions to the L.P.P. Also suppose that x is a basic feasible solution

$$x = \begin{bmatrix} x_B \\ 0 \end{bmatrix}$$

where x_B is an $m \times 1$ vector, such that for a none-singular submatrix of B of A

$$Bx_B = b$$

If possible, let x be a point of S, such that there exist $x_1, x_2 \in S$, so that

$$x = \lambda x_1 + (1 - \lambda) x_2, 0 < \lambda < 1$$

Let $x_1 = \begin{bmatrix} u_1 \\ v_1 \end{bmatrix}$ and $x_2 = \begin{bmatrix} u_2 \\ v_2 \end{bmatrix}$

where $u_1\, u_2$ are $m \times 1$ vectors and v_1, v_2 are $(n-m) \times 1$ Vectors.

Then $\begin{bmatrix} x_B \\ 0 \end{bmatrix} = \lambda \begin{bmatrix} u_1 \\ v_1 \end{bmatrix} + (1-\lambda) \begin{bmatrix} u_2 \\ v_2 \end{bmatrix}$

and $0 = \lambda v_1 + (1-\lambda)\, v_2$

Since x_1, x_2 are feasible solutions, therefore,

$$u_1, u_2, v_1, v_2 \geq 0$$

Now, $0 < \lambda < 1$ and $0 = \lambda v_1 + (1-\lambda)\, v_2$

$\therefore$ we must have

$$v_1 = v_2 = 0$$

Thus, $x_1 = \begin{bmatrix} u_1 \\ 0 \end{bmatrix}, x_2 = \begin{bmatrix} u_2 \\ 0 \end{bmatrix}$

Again since x_1, x_2 satisfy

$$Ax_1 = b \text{ and } Ax_2 = b$$

therefore,

$$Bu_1 = b \text{ and } Bu_2 = b$$

Also since $BX_B = b$ and since expression of b as a linear combination of basis vectors must be unique, therefore

$$u_1 = u_2 = x_B$$

Hence, $x = x_1 = x_2$

This is a contradiction, for by assumption $x_1 \neq x_2$

Hence x is an extreme point of S.

Theorem: *(Reduction of a feasible solution to a Basic Feasible Solution)*. If an L.P.P. has a feasible solution, then it also has a basic feasible solution.

Proof: Let the L.P.P. be to determine x so as to maximize

$$Z = cx, \quad x^T \in R^n$$

Subject to the constraints:

$$Ax = b$$

$$x \geq 0$$

where A is an $m \times n$ real matrix and b, c are $m \times 1$ and $1 \times n$ real matrices respectively. Let $\rho(A) = m$ i.e. rank of A is m.

Since there does not exist a feasible solution, we have

$$\rho(A, b) = \rho(A) \text{ and } m < n.$$

Let $x = (x_1, x_2, \ldots, x_n)$ be feasible solution so that $x_j \geq 0$ for all j.

To be precise let us suppose that x has p positive components and let the remaining $n - p$ components be zero.

Let us so relabel our components that the positive components are the first p components and assume that the columns of A have been relabelled accordingly,

Then $$\sum_{j=1}^{p} a_j x_j = b$$

where $a_1, a_2, \ldots a_p$ are first p columns of A.

Two cases now do arise:

(i) The vectors $a_1, a_2, \ldots a_p$ form a linearly independent set. Then $p \leq m$.

If $p = m$; the given solution is a non-degenerate basic feasible solution with $x_1, x_2 \ldots, x_p$ as basic variables.

If $p < m$, then the set $\{a_1, a_2, \ldots, a_p\}$ can be extended to $\{a_1, a_2, \ldots a_p, a_{p+1}, \ldots a_m)$ to form a basis for the columns of A. Then we have

$$\sum_{j=1}^{m} a_j x_j = b$$

where $x_j = 0$ for $j = p + 1, p + 2, \ldots, m$.

Thus, we have, in this case, a degenerate basic feasible solution with $m - p$ of basic variables zero.

(ii) The set $\{a_1, a_2, \ldots, a_p)$ is linearly dependent obviously $p > m$. Let $\{a_1\ a_2, \ldots, a_p\}$ be a set of constants (not all zero) such that

$$\sum_{j=1}^{p} a_j x_j = 0$$

Suppose that for any index r, $a_r \neq 0$. Then

$$a_r = -\sum_{\substack{j \neq r \\ j=1}}^{p} \frac{a_j}{a_r} a_j$$

Substituting in the relation

$$\sum_{j=1}^{p} a_j x_j = b$$

we get

$$\sum_{\substack{j \neq r \\ j=1}}^{p} a_j x_j + \left(-\sum_{\substack{j \neq r \\ j=1}}^{p} \frac{a_j}{a_r} a_j \right) x_r = b$$

$$\sum_{\substack{j \neq r \\ j=1}}^{p} \left(x_j - x_r \frac{a_j}{a_r} \right) a_j = b$$

Thus we have a solution with no more than $p - 1$ non-zero components. So ensure that these are non-negative we shall choose a_r in such a way that

$$x_j - x_r \frac{a_j}{a_r} \geq 0 \text{ for all } j \neq r$$

This requires that either $a_j = 0$ or

$$\frac{x_i}{a_r} \geq \frac{x_r}{a_r} \text{ if } a_j > 0$$

and $$\frac{x_i}{a_j} \leq \frac{x_r}{a_r} \text{ if } a_j < 0$$

Thus, if we select a_r such that

$$\frac{x_r}{a_j} = \frac{\min}{j} \left\{ \frac{x_i}{a_j} \quad a_j > 0 \right\}$$

then each of the $p - 1$ variables $x_j - x_r \dfrac{a_j}{a_r}$ a non-negative, and so we have a feasible solution with not more than $p - 1$ non-zero component.

Consider now this new feasible solution with not more than $p - 1$ non-zero components. If the corresponding set of $p - 1$ columns of A is linearly independent, case (1) applies and we have arrived at a basic feasible solution. If this set is again linearly dependent, we may repeat the process to arrive at a feasible solution with not more than $p - 2$ non-zero components. The argument can be repeated. Ultimately, we get a feasible solution with set of column vectors of A-linearly independent. The discussion of case (i) applies and we do get a basic feasible solution.

This completes the proof.

Examples

Example 1. If $x_1 = 2$, $x_2 = 3$, $x_3 = 1$ be a feasible solution of linear programming problem.

Max. $\quad z = x_1 + 2x_2 + 4x_3$

Subject to $\quad 2x_1 + x_2 + 4x_3 = 11$

$3x_1 + x_2 + 5x_3 = 14$

$x_1, x_2, x_3 \geq 0$

then find a basic feasible solution.

[Meerut '70, '77; Punjab (M.B.A.) '77]

Solution: We express the above system as

$$\begin{matrix} a_1 & a_2 & a_3 \end{matrix} \quad x \quad\quad b$$

$$\begin{bmatrix} 2 & 1 & 4 \\ 3 & 1 & 5 \end{bmatrix} \begin{bmatrix} x_1 \\ x_2 \\ x_3 \end{bmatrix} = \begin{bmatrix} 11 \\ 14 \end{bmatrix}$$

or $x_1 a_1 + x_2 a_2 + x_3 a_3 = b$

But we are given that the feasible solution is $x_1 = 2, x_2 = 3, x_3 = 1$

Hence

$$2a_1 + 3a_2 + 1a_3 = b$$

where $a_1 = \begin{bmatrix} 2 \\ 3 \end{bmatrix}, a_2 = \begin{bmatrix} 1 \\ 1 \end{bmatrix}$

$$a_3 = \begin{bmatrix} 4 \\ 5 \end{bmatrix} \quad b = \begin{bmatrix} 11 \\ 14 \end{bmatrix}$$

Since the vectors a_1, a_2, a_3 associated with the corresponding variables x_1, x_2, x_3 are L.D. (linearly dependent), therefore one of the vector can be expressed in terms of the remaining two.

Thus,

$$a_3 = \lambda_1 a_1 + \lambda_2 a_2 \tag{1}$$

or

$$\begin{bmatrix} 4 \\ 5 \end{bmatrix} = \lambda_1 \begin{bmatrix} 2 \\ 3 \end{bmatrix} + \lambda_2 \begin{bmatrix} 1 \\ 1 \end{bmatrix}$$

$$= \begin{bmatrix} 2\lambda_1 \\ 3\lambda_1 \end{bmatrix} + \begin{bmatrix} \lambda_1 \\ \lambda_2 \end{bmatrix}$$

$$= \begin{bmatrix} 2\lambda_1 + \lambda_2 \\ 3\lambda_1 + \lambda_2 \end{bmatrix}$$

which gives

$$2\lambda_1 + \lambda_2 = 4$$

$$3\lambda_1 + \lambda_2 = 5$$

Solving these two equations, we have $\lambda_1 = 1$, $\lambda_2 = 2$. Putting these values of λ_1 and λ_2 in (1) we get.

$$a_1 + 2a_2 - a_3 = 0 \tag{2}$$

which we put as $\lambda_1 a_1 + \lambda_2 a_2 + \lambda_3 a_3 = 0$

where $\lambda_1 = 1, \lambda_2 = 2, \lambda_3 = -1$

Now, we determine which of the three variables (x_1, x_2, x_3) should be zero.

First, we find

$$v = \max\left(\frac{\lambda_1}{x_1}, \frac{\lambda_2}{x_2}, \frac{\lambda_3}{x_3}\right)$$

$$= \max\left(\frac{1}{2}, \frac{2}{3}, \frac{-1}{1}\right)$$

$(\because \lambda_1 = 1, \lambda_2 = 2, \lambda_3 = -1)$

and given that

$$x_1 = 2$$
$$x_2 = 3$$
$$x_3 = 1$$
$$\therefore v = \frac{2}{3}$$

Since $\left(x_i = \frac{\lambda_i}{v}\right) \geq 0$, therefore

$$x^* = \left(x_1 - \frac{\lambda_1}{v}, x_2 - \frac{\lambda_2}{v}, x_3 - \frac{\lambda_3}{v}\right)$$

will be a reduced so lution.

But

$$x_1 - \frac{\lambda_1}{v} = 2 - \frac{1}{\frac{2}{3}} = \frac{1}{2}$$

$$x_2 - \frac{\lambda_2}{v} = 3 - \frac{2}{\frac{2}{3}} = 0$$

$$x_3 - \frac{\lambda_3}{v} = 1 - \left(-\frac{1}{\frac{2}{3}}\right) = \frac{5}{2}$$

Now, this solution

$x = \left(\frac{1}{2}, 0, \frac{5}{2}\right)$ must be basic feasible if the vectors.

$$a_1 = \begin{bmatrix} 2 \\ 3 \end{bmatrix}$$

and

$$a_3 = \begin{bmatrix} 4 \\ 5 \end{bmatrix}$$

associated with non-zero variables x_1 and x_2 are linearly independent. Clearly a_1 and a_2 are linearly independent. Hence, the required basic feasible solution is

$$x_1 = \frac{1}{2}, x_2 = 0, x_3 = \frac{5}{2}.$$

Example 2. If $x_1 = 2, x_2 = 4, x_3 = 1$ be a feasible solution of linear equation

$$2x_1 - x_2 + 2x_3 = 2$$
$$x_1 + 4x_2 = 18$$
$$x_1, x_2 \geq 0$$

then find a B.F.S. (basic feasible solution).

Solution: The given equations can be expressed as

$$\overset{a_1,\ a_2,\ a_3}{\begin{bmatrix} 2 & -1 & 2 \\ 1 & 4 & 0 \end{bmatrix}} \overset{x}{\begin{bmatrix} x_1 \\ x_2 \\ x_3 \end{bmatrix}} = \overset{b}{\begin{bmatrix} 2 \\ 18 \end{bmatrix}}$$

or $x_1 a_1 + x_2 a_2 + x_3 a_3 = b$

But the feasible solution is given as

$x_1 = 2,\ x_2 = 4,\ x_3 = 1$

Hence, $2a_1 + 4a_2 + 1a_3 = b$

where $$a_1 = \begin{bmatrix} 2 \\ 1 \end{bmatrix},\quad a_2 = \begin{bmatrix} -1 \\ 4 \end{bmatrix}$$

$$a_3 = \begin{bmatrix} 2 \\ 0 \end{bmatrix},\quad b = \begin{bmatrix} 2 \\ 18 \end{bmatrix}$$

Since the vectors a_1, a_2, a_3 associated with the corresponding variables x_1, x_2, x_3 are linearly dependent, therefore one of the vectors can be expressed in terms of the remaining two

Thus $$a_3 = \lambda_1 a_1 + \lambda_2 a_2 \qquad ...(1)$$

or $$\begin{bmatrix} 2 \\ 0 \end{bmatrix} = \lambda_1 \begin{bmatrix} 2 \\ 1 \end{bmatrix} + \lambda_2 \begin{bmatrix} -1 \\ 4 \end{bmatrix}$$

$$= \begin{bmatrix} 2\lambda_1 \\ \lambda_1 \end{bmatrix} + \begin{bmatrix} -\lambda_2 \\ 4\lambda_2 \end{bmatrix}$$

$$= \begin{bmatrix} 2\lambda_1 - \lambda_2 \\ \lambda_1 + 4\lambda_2 \end{bmatrix}$$

$$\therefore\ 2\lambda_1 - \lambda_2 = 2$$

$$\lambda_1 + 4\lambda_2 = 0$$

Solving $$\lambda_1 = \frac{8}{9},\quad \lambda_2 = -\frac{2}{9}$$

So (1) becomes

$$a_3 = \frac{8}{9} a_1 + \left(\frac{-2}{9}\right) a_2$$

or $$a_3 = \frac{8}{9} a_1 + -\frac{2}{9} a_2 + (-1) a_3 = 0$$

which is of the form

$$\lambda_1 a_1 + \lambda_2 a_2 + \lambda_3 a_3 = 0$$

where $\lambda_1 = \dfrac{8}{9},\ \lambda_2 = -\dfrac{2}{9},\ \lambda_3 = -1$

Now we determine which one of the three variables (x_1, x_2, x_3) should be 0.

Now, let $$V = \max\left(\frac{\lambda_1}{x_1}, \frac{\lambda_2}{x_2}, \frac{\lambda_3}{x_3}\right)$$

$$= \max\left(\frac{\frac{8}{9}}{2}, \frac{\frac{2}{9}}{4}, \frac{-1}{1}\right)$$

$$= \max\left(\frac{4}{9}, \frac{-1}{18}, -1\right)$$

$$V = \frac{4}{9}$$

Since $\left(x_i - \frac{\lambda_i}{v}\right) \geq 0$, therefore

$$x^* = \left(x_1 - \frac{\lambda_1}{v}, x_2 - \frac{\lambda_2}{v}, x_3 - \frac{\lambda_3}{v}\right)$$

will be reduced solution

$$x_1 - \frac{\lambda_1}{v} = 2 - \frac{\frac{8}{9}}{\frac{4}{9}} = 2 - 2 = 0$$

$$x_2 - \frac{\lambda_2}{v} = 4 - \left(\frac{\frac{-2}{9}}{\frac{4}{9}}\right) = 4 + \frac{1}{2} = \frac{9}{2}$$

$$x_3 - \frac{\lambda_3}{v} = 1 - \frac{(-1)}{\frac{4}{9}} = 1 + \frac{9}{4} = \frac{13}{4}$$

Now, this solution

$x^* = \left(0, \frac{9}{2}, \frac{13}{4}\right)$ must be a basic feasible if the vectors

$$a_2 = \begin{bmatrix} -1 \\ 4 \end{bmatrix} \quad \text{and} \quad a_3 = \begin{bmatrix} 2 \\ 0 \end{bmatrix}$$

associated with non-zero variables x_2 and x_3 are linearly independent clearly vectors a_2 and a_3 are linearly independent hence the required basic feasible solution is

$$x_1 = 0, \quad x_2 = \frac{9}{2}, \quad x_3 = \frac{13}{4}$$

Example 3. Show that the feasible solution

$x_1 = 1, x_2 = 0, x_3 = 1$ to the system

Min. $z = 2x_1 + 3x_2 + 4x_3$

s.t. $x_1 + x_2 + x_3 = 2$

$x_1 - x_2 + x_3 = 3$ is not basic.

Solution: Here we express the constraints as

$$A \quad x \quad b$$

$$= \begin{bmatrix} 2 \\ 3 \end{bmatrix}$$

or $$Ax = \begin{bmatrix} a_1 & a_2 & a_3 \\ 1 & 1 & 1 \\ 1 & -1 & 1 \end{bmatrix} \begin{bmatrix} x_1 \\ x_2 \\ x_3 \end{bmatrix}$$

where $$A = \begin{bmatrix} 1 & 1 & 1 \\ 1 & -1 & 1 \end{bmatrix}$$

$$x = \begin{bmatrix} x_1 \\ x_2 \\ x_3 \end{bmatrix}$$

$$b = \begin{bmatrix} 2 \\ 3 \end{bmatrix}$$

Also $$a_1 = \begin{bmatrix} 1 \\ 1 \end{bmatrix}, \quad a_2 = \begin{bmatrix} 1 \\ -1 \end{bmatrix}, \quad a_3 = \begin{bmatrix} 1 \\ 1 \end{bmatrix}$$

here $$a_1 = \begin{bmatrix} 1 \\ 1 \end{bmatrix} \quad \text{and} \quad a_3 = \begin{bmatrix} 1 \\ 1 \end{bmatrix}$$

Since there exists non-zero scalar $\lambda_1 = 1, \lambda_2 = -1$ such that

$$\lambda_1 a_1 + \lambda_2 a_3 = 0$$

or $$1a_1 + (-1)\, a_3 = 0$$

or $$1 \begin{bmatrix} 1 \\ 1 \end{bmatrix} + (-1) \begin{bmatrix} 1 \\ 1 \end{bmatrix} = 0$$

or $$\begin{bmatrix} 1 \\ 1 \end{bmatrix} + \begin{bmatrix} -1 \\ -1 \end{bmatrix} = \begin{bmatrix} 0 \\ 0 \end{bmatrix}$$

or $$\begin{bmatrix} 1 & -1 \\ 1 & -1 \end{bmatrix} = \begin{bmatrix} 0 \\ 0 \end{bmatrix}$$

which is true. Hence vectors a_1 and a_3 are linearly dependent we know that the any feasible solution to $Ax = b$ for which vectors $a_1, a_2, a_3 ..., a_n$ associated with the non zero variables $x_1, x_2, x_3 ... x_n$ respectively are linearly independent, is a basic feasible solution.

But here vectors a_1 and a_3 are linearly dependent.

Hence the given feasible solution is not basic.

Example 4. The column vector [1, 1, 1] is a feasible solution to the system of equations

$$x_1 + x_2 + 2x_3 = 4$$
$$2x_1 - x_2 + x_3 = 2$$

Reduce the given feasible solution to basic feasible solution.

Solution: The given system of equations can be written as

$$\begin{pmatrix} 1 & 1 & 2 \\ 2 & -1 & 1 \end{pmatrix} \begin{bmatrix} x_1 \\ x_2 \\ x_3 \end{bmatrix} = \begin{pmatrix} 4 \\ 2 \end{pmatrix}$$

or $$Ax = b$$

Since $p(A) = 2$, therefore there are only two linearly independent column vectors of A, let the column vectors of A be denoted by a_1, a_2 and a_3 respectively.

Thus,

(i) $\alpha_1 a_1 + \alpha_2 a_2 + \alpha_3 a_3 = 0$, (1) a_i are scalars not of all zero we must have

$$a_1 + a_2 + 2a_3 = 0$$

and $$2a_1 - a_2 + \alpha_3 = 0$$

there give $$a_1 = -\alpha_3$$

Taking $a_3 = 1$, we get $a_1 = -1$ and $a_2 = -1$. The equation of then becomes

$$-\alpha_1 - \alpha_2 + \alpha_3 = 0$$

The given system of equation is

$$x_1 a_1 + x_2 a_2 + x_3 a_3 = b$$

where $$x_1 = x_2 = x_3 = 1$$

In order to reduce the number of positive variables, the vector to be removed is selected

$$\text{Now } \frac{x_r}{a_r} = \underset{j}{Min} \left\{ \frac{x_i}{a_j}, aj > 0 \right\} = \underset{j}{Min} \begin{Bmatrix} 1 \\ 1 \end{Bmatrix} 1, j = 1, 2, 3$$

The vector for which $\frac{x_r}{a_r} = 1$, is a_3 and so we can remove the vector a_3 and obtain a new solution with not more than two non-negative variables thus the values of the new variables are given by

$$\hat{x}_j = x_j - \frac{x_r}{a_r} a_j$$

this gives
$$\hat{x}_1 = x_1 - \frac{x_r}{a_r} a_1 = 1 - 1(-1) = 2$$

$$\hat{x}_2 = x_2 - \frac{x_r}{a_r} a_2 = 1 - 1(-1) = 2$$

Thus [2, 2, 0] is also a feasible solution, and it has only two non-zero components and so

$2a_1 + 2a_2 = b$

(iv) $x_1 + 2x_2 - 9x_3 - 10x_4 = -5$

$x_1 + x_2 + 3x_3 + 3x_4 = 6$

$x_1 + 2x_2 - 3x_3 - 4x_4 = 1$

Ans. (1, 2, 0, 1), (2, 1, 1, 0)

Example 5. Consider the system of equations

$$x_1 + 2x_2 + 4x_3 + x_4 = 7$$
$$2x_1 - x_2 + 3x_3 - 2x_4 = 4$$

Here $x_1 = 1, x_2 = 1, x_3 = 1$ and $x_4 = 0$ is a feasible solution. Reduce this feasible solution to two different basic feasible solutions.

Ans. $\left(0, \frac{1}{2}, 3/2, 0\right)$; (3, 2, 0, 0)

Theorem : (Minimax theorem): Let f be a linear function of n variables such that $f(x^*)$ is its minimum value for some point x^{T*}, $x^T \in R^n$. Then, $-f(x)$ attains its maximum at the point x^*.

Moreover, for $x^T \in R^n$

Minimum $f(x) = -$Maximum $\{-f(x)\}$.

Proof: Since $f(x)$ is a minimum at the point x^*, therefore,

$$f(x^*) \le f(x) \qquad \forall\, x^T \in R^n$$

or,
$$-f(x^*) \ge -f(x) \qquad \forall\, x^T \in R^n$$

This shows that $-f(x)$ attains its maximum at point x^*.

Thus,

$$\text{Maximum } \{-f(x)\} = -f(x^*)$$

Again, since minimum $f(x) = f(x^*)$ [given] therefore,

$$\text{Minimum } f(x) = f(x^*) = -\text{ Maximum } \{-f(x)\}$$

Hence,

$$\text{Minimum } f(x) = -\text{ Maximum } \{-f(x)\} \text{ for } x^T \in R^n.$$

Example : Let us verify the above theorem for the function f which assumes the values

$$f(x) = (9, 7, 5, 3, 1) \text{ for } x^T \in R^n.$$

Now since a_1, a_2 are linearly independent, [2, 2, 0] is a basic feasible solution, with $x_3 = 0$ as a non-basic variable.

Note: The choice of a_i^s in the relation

$$\alpha_1 a_1 + \alpha_2 a_2 + \alpha_3 a_3 = 0$$

has a bearing 0 the basic feasible solution obtained. For example, if we take

$$\alpha_1 = 1,\ \alpha_3 = -1 \text{ and } \alpha_2 = 1, \text{ so that}$$
$$a_1 + a_2 - a_3 = 0$$

then $$\underset{j}{\text{Min}} \left\{ \frac{x_1}{\alpha_j}, \alpha_j > 0 \right\} = \text{Min} \left\{ \frac{1}{1}, \frac{1}{1} \right\} = 1.$$

Here $$\frac{x_1}{\alpha_1} = \frac{x_2}{\alpha_2} = 1 = \underset{j}{\text{Min}} \left\{ \frac{x_1}{\alpha_j}, \alpha_j > 0 \right\}$$

thus a_1 and a_2 are equally eligible for removal.
In case a_1 is removed, we get

$$\hat{x}_2 = x_2 - \frac{x_r}{\alpha_r} a_2 = 1 - 1 = 0$$

$$\hat{x}_3 = x_3 - \frac{x_r}{\alpha_r} a_3 = 1 - (-1) = 2$$

and in case a_2 is removed, we get

$$\hat{x}_1 = x_r - \frac{x_r}{\alpha_r} \alpha_1 = 1 - 1 = 0$$

$$\hat{x}_3 = x_3 - \frac{x_r}{\alpha_r} \alpha_2 = 1 - (-1) = 2$$

It is shown, as before, that the two solutions are basic.

Hence corresponding to the choice of α_j^s made here, we get two basic feasible solution (both degenerate):

(i) [0, 0, 2] with $x_1 = 0$, non-basic.
and (ii) [0, 0, 2] with $x_2 = 0$, non-basic.

EXERCISE

1. Find out all the basic solutions of the system

$$x_1 - x_2 + 2x_3 = 18$$
$$x_1 + 2x_2 - x_3 = 1$$

and show that one of them is not feasible.

Ans. (14, 0, 2), (0, 14, 16) and (16, –2, 0).

2. Show that the following system

$$2x_1 - 4x_2 - 2x_3 - 5x_4 = 2$$
$$10x_1 + 8x_2 + 6x_3 - 7x_4 = 10$$

has only one basic feasible solution.

Ans. (1, 0, 0)

3. Find all the basic feasible solutions of the system

$$2x_1 + x_2 - x_3 + x_5 = 14$$
$$x_1 + 3x_2 + x_4 = 15$$

Ans. $x_1 = \frac{27}{5}, x_2 = \frac{16}{5}, x_1 = 15, x_3 = 16; x_1 = 7, x_4 = 8, x_2 = 5,$ $x_5 = 9$ and $x_4 = 15, x_5 = 14.$

4. Find all the basic feasible solutions of the following:

(i) $x_1 - 2x_2 - x_3 = 10$

$2x_1 + x_2 + 2x_3 = 12$

Ans. None

(ii) $4x_1 + 2x_2 + 6x_3 + 8x_4 = 3$

$6x_1 + 8x_2 + 4x_3 + 2x_4 = 7$

Ans. $x_1 = \frac{1}{2}; x_2 = \frac{1}{2}; x_2 = \frac{3}{4}; x_3 = \frac{1}{4}; x_2 = \frac{5}{6}; x_4 = \frac{1}{6}$

(iii) $2x_1 + 6x_2 + 2x_3 + x_4 = 3$

$6x_1 + 4x_2 + 4x_3 + 6x_4 = 2$

Ans. $x_2 = \frac{1}{2}$ i.e. $\left(0, \frac{1}{2}, 0, 0\right)$

We have, $-f(x) = (-9, -7, -5, -3, 1)$

$\therefore$ $\quad$ Maximum $\{-f(x)\} = -1$

or $\quad$ $-$Maximum $\{(-f(x)\} = 1$

Also, $\quad$ Minimum $\{f(x)\} = 1$

Thus, $\quad$ Minimum $f(x) = -$Maximum $\{-f(x)\}$.

Simplex Algorithm for Maximization L.P.P. [R.U.M.Sc. '83]

Computational Procedure—It consists of the following steps systematically.

Step I: If the problem is of minimization, change it into the maximizations problem. For an example if the problem is

$$\text{Minimize } Z = c_1x_1 + c_2x_2 + \ldots + c_nx_n$$

Then put it as

$$\text{Maximize } Z^* = (-Z) = -c_1x_1 - c_2x_2 \ldots - c_nx_n$$

If $Z^* = M$ be the maximum value of Z^*.

then $-M$ will be the minimum value of Z.

Step II: Make all bi's positive.

If any of the bi's is negative multiply the corresponding constraint by -1.

Let the constraint be

$$2x_1 - x_2 + x_3 \leq -6$$

Here $b = -6$ is negative. Hence multiplying both sides by -1,

$$-2x_1 + x_2 - x_3 \geq 6$$

Step III: Convert the constraints into equations by introducing the non-negative slack or Surplus Variables.

Also, introduce artificial variables in the constraints where surplus variables are added.

Step IV: To obtain Starting Basic Feasible Solution (B.F.S.). This can be done as follows.

(a) *When all the original constraints have ≤ signs. Put all the non-basic variables (i.e. given variables)* $x_1, x_2, \ldots x_n$ equal to zero and solve the equations for the slack variables.

Let the constraint be

$$x_1 + 2x_2 \leq 5$$

By adding slack variables, we get

$$x_1 + 2x_2 + x_3 = 5$$

Putting $x_1 = 0$, $x_2 = 0$, we get $x_3 = 5$

(b) When all the original constraints have ≥ signs.

Put all the non-basic variables (i.e. given variables and surplus variables) equal to zero and solve the equations for basic artificial variables.

Let the constraint be

$$5x_1 + x_2 \geq 6$$

Substract surplus variable x_3 to make it an equation. Then we have

$$5x_1 + x_2 - x_3 = 6$$

By putting $x_1 = 0, x_2 = 0$, we get $x_3 = -6$ which is not B.F.S. as it is negative.

So again, we have to add a new variable called Artificial variable, say x_4. Then we get

$$5x_1 + x_2 - x_3 + x_4 = 6$$

Put the given variable (x_1, x_2) and the surplus variable x_3 equal to zero. Then the artificial variable $x_4 = 6$ which is B.F.S. as it is +ve.

(c) When the constraints have ≤, ≥ and = signs.

In this case the constraints are converted into equations by putting slack. Surplus and artificial variables.

Put all the non-basic variables equal to zero and solve for remaining basic variables.

Step V: Construction of starting simplex table.

B (Program) (Slack variables introduced)	*C_B Co-eff. of slack variables in objective function*	*X_B (Quantity) Magnitude of slack variables after putting $x_1 = x_2 = \ldots = x_n = 0$*	*$Y_1(-a_1)$ Co-eff. of x_1 in constraints*
Y_{n+1} Y_{n+1} Y_{n+r} Y_{n+m}	$C_{B_1} = 0$ $C_{B_2} = 0$ $C_{B_r} = 0$ $C_{B_m} = 0$	$X_{B_1} = b_1$ $X_{B_2} = b_2$ $X_{B_r} = br$ $X_{B_m} = bm$	$y_{11} = a_{11}$ $y_{21} = a_{21}$ $y_{r_1} = a_{r_1}$ $y_{m1} = a_{m1}$
$Z = C_B X_B$ $= 0 \times b_1 + 0 \times b_2 + \ldots + 0 \times b_m$ $= 0$		x_j = (row of starting B.F.S.) i.e. value of starting B.F.S.	$x_1 = 0$
		C_j = Co-eff. of given variable and slack variables in objective function.	c_1
		Δ_j (Net evaluation row or Basic row) $= C_j - C_B Y_j$	Δ_r

$Y_2 (= a_2)$ Co-eff. of x_2 in constraints	…	*$Y_k (= a_k)$*	*$Y_n (= a_n)$ Co-eff. of x_n in constraints*	*$Y_{n+1} (= \beta_1)$ Co-eff. of slack variable constraints*	…	*$Y_{n+m} (= \beta_m)$ co-eff. of slack vartable y_{n+m} in const-raints*	*minimum ratio XB — Y_k*
$y_1 = a_{12}$ $y_{22} = a_{22}$ ……… $y_{t2} = a_{t2}$ $y_{m2} = a_{m2}$	 … 	$y_1k = a_1k$ $y_2k = a_2k$ ……… □ $y_{mk} = a_{mk}$	$y_1n = a_1n$ $y_2n = a_2n$ ……… $y_{rn} = a_{rn}$ $a_{nm} = a_{mn}$	1 0 ……… 0 0	 … 	0 0 ……… 0 1	 ⇒
$x_2 =$	…	$x_k = 0$	$x_n = 0$	$x_{n+1} = b_1$	…	$x_{n+m} = b_n$	
c_2	…	c_k	c_n	0	…	0	
Δ_2	…	Δ_k ↑	Δ_n	Δ_{n+1}	…	Δ_{n+m}	

Note: Slack variables $y_{n+1}, y_{n+2}, \ldots y_{n+m}$ are also denoted by $S_{n+1}, S_{n+2}, \ldots S_{n+m}$.

***Step VI*:** Test the starting B.F.S. for optimality.

(a) If $\Delta_j \leq 0$ for each j, then the solution under test is optimal.

(i) If one of Δ_j is +ve, but any are zero, then other optimal solution exist with the same value of Z.

(ii) If all of Δ_j are negative, solution under test is unique optimal solution.

(b) if $\Delta_j > 0$ for any j i.e., of one or more of Δ_j are +ve, the solution under test is not optimal.

If the solution under test is not optimal, we proceed to the next step.

Step VII: (i) To find incoming (or, entering or key column) and out going (or key row). Locate the largest positive number in the net evaluation row i.e. (locate the maximum value of Δ). The column under which this number lies is called key column (or incoming denoted by ↑).

If maximum value of Δ occurs at more than one, then any one of them can be taken as an incoming vector.

(ii) To find outgoing vector (or key row).

The outgoing vector β_r, is taken corresponding to that value of r for which

$$\frac{x_{\beta_r}}{y_{rk}} = \underset{i}{\text{Mini}}\left\{\frac{x_{Bi}}{y_{ik}}, y_{ik} > 0\right\}$$

where a_k is the incoming vector.

This can also be put in the following way.

Obtain the entries under the column (Minimum ratio) by dividing the entries under the column X_B (for Quantity) by the corresponding non-negative entries of the key column. The row in which the smallest ratio lies is called key row denoted by $\rightarrow$.

Step VIII: Key element (or Pivot element). The number which lies at the intersection of the key column and key row is called the key number. We denote this element by □.

Step IX: Transformation of key row and non-key row.

Let a_k be the incoming vector and β_r the outgoing. Then the element ($y_{rk} = a_{rk}$) is called the key element or pivot element.

In order to bring β_r in place of a_k (= Y_k) there should be unity at the key element. If it is not 1, then divide all the elements of these row by this key element. Then subtract appropriate multiples of this row (containing key element) from all the other rows and obtain zero at all other positions of this column a_k (= Y_k). Now bring β_r in place a_k (= Y_k) and construct new (revised) simplex table.

Calculations of key row and non-key row can also be made as follows.

If the key column falls under the variable Y_k replace the corresponding variable in column B by that variable Y_k and put value of Y_k (i.e. co-eff. Y_k in objective function in the column C_B.

Transformation of key row—Divide all the numbers in the key row by the key number.

Transformation of non-key row—The transformation of non-key row can be made by the following rule.

New row number = Old row number – (corresponding number in key row) × corresponding fixed ratio.

$$\text{where fixed ratio} = \frac{\text{Old row number in key column}}{\text{Key number}}$$

In this way we get improved basic feasible solution.

Step X: Now, test the above B.F.S. for optimality as in Step 6. In this solution is not optimal, then repeat (7) and (8) until and optimal solution is finally obtained.

For clear understanding of the above procedure some problems specially examples (1 and 3) have been worked out in detail in two different ways.

FLOW CHART OF SIMPLEX METHOD

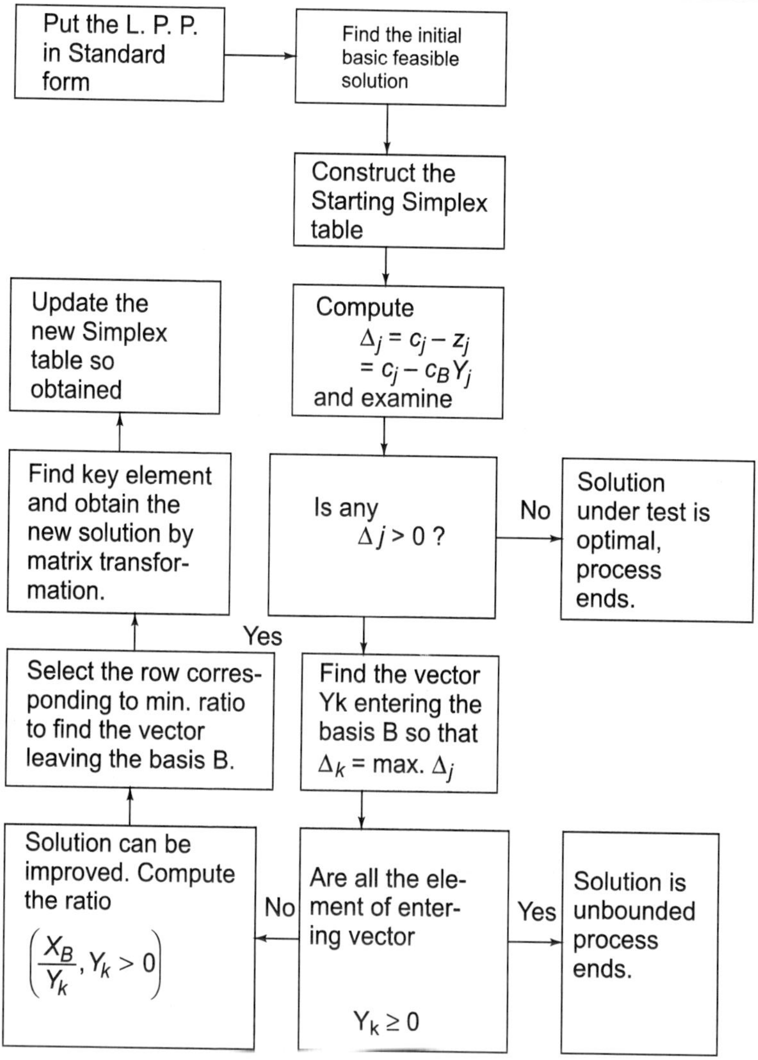

Illustrative Examples

1. Solve the following L.P.P. by simplex method,

Max. $Z = 3x_1 + 2x_2$

Subject to

$$x_1 + x_2 \leq 4$$
$$x_1 - x_2 \leq 2$$
$$x_1, x_2 \geq 0 \qquad \text{[R.U., B.Sc. '88]}$$

Here $c_1 = 3$, $c_2 = 2$; $b_1 = 4$, $b_2 = 2$

[V.B.U. Part III Hons. '99]

Solution:

Step 1. The problem is a problem of maximization.

Step 2. All the bi's (here b_1 and b_2) are positive.

Step 3. Now the inequalities are converted into equations by introducing slack variables x_3 and x_4 as follows

$$x_1 + x_2 + x_3 = 4$$

and $$x_1 - x_2 + x_4 = 2$$

These two equations can be written as

$$x_1 + x_2 + x_3 + 0x_4 = 4 \quad \text{... (i)}$$

and $$x_1 - x_2 + 0x_3 + x_4 = 2 \quad \text{... (ii)}$$

The objective function

$$Z = 3x_1 + 2x_2$$

can be put as

$$Z = 3x_1 + 2x_2 + 0x_3 + 0x_4 \quad \text{... (iii)}$$

From this we have

$$c_1 = 3, c_2 = 2, c_3 = 0, c_4 = 0$$

Step 4. Putting the decision variables (that is, x_1 and x_2) each equal to zero in (i) and (ii), we get $x_3 = 4$ and $x_4 = 2$.

Thus starting B.F.S. (basis feasible solution) is

$$x_1 = 0, x_2 = 0, x_3 = 4, x_4 = 2$$

Step 5. Construct the starting simplex table 2

Step 6. Compute Δ_j by formula

$$\Delta_j = c_j - C_B Y_j$$

$$\therefore \quad \Delta_1 = c_1 - C_B Y_1$$

$$= 3 - (0, 0)(1, 1)$$

$$= 3 - (0 \times 1 + 0 \times 1) = 3$$

$$= 2 - (0, 0)(1, 1) = 2 - 0 = 2$$

$$\Delta_3 = c_3 - C_B Y_3$$

$$= 0 - (0, 0)(1, 0) = 0$$

$$\Delta_4 = c_4 - C_B Y_4$$

$$= 0 - (0, 0)(0, 1) = 0$$

Since all Δ_j's are not less than or equal to zero, therefore the solution is not optimal. So we proceed to next step.

TABLE 1

B Slack variables introduced	C_B Coeff. of slack variables in objective function (iii) on (iii)	X_B Value of slack variables after putting $x_1 = 0$, $x_2 = 0$	$Y_1\ (= a_1)$ Coeff. of x_1 in constraints	$Y_2\ (= a_2)$ Coeff. of x_2 in constraints	$Y_3\ (= \beta_1)$ Coeff. of slack variables x_3 in constaints	$Y_4\ (= \beta_2)$ Coeff. slack variables x_4 in constaints	Minimum ratio $= X_B$ Incoming vector $= \frac{X_B}{Y_1}$
Y_3	0 = coeff. of x_3 in objective function	$x_3 = 4$	1	1	1	0	$\frac{4}{1} = 4$
Y_4	0 = coeff. of x_4 in objective function	$x_4 = 2$	$\boxed{1}$	– 1	0	1	$\frac{2}{1} = 2$ →
$Z = C_B X_B$		x_j	$x_1 = 0$	$x_2 = 0$	$x_3 = 4$	$x_4 = 2$	
$= 0 \times 4 + 0 \times 2$		c_j	$c_1 = 3$	$c_2 = 2$	$c_3 = 0$	$c_4 = 0$	
$= 0$		Δ_j	$\Delta_1 = 3$	$\Delta_2 = 2$	$\Delta_3 = 0$	$\Delta_4 = 0$	
			↑ incoming vector			↓ outgoing vector	

Step 7: To find incoming and outgoing vector. The largest value of Δ is 3 which lies in the column Y_1 (= a_1). Therefore y_1 (a_1) is incoming vector.

Since the minimum ratio $\left(\frac{2}{1} = 2\right)$ occurs in the row Y_4, the column Y_4 = (= β_2) is outgoing vector. Hence in the next improved table write Y_1 in place of Y_4 of column B (i.e. first column) and replace "0 (= coeff of x_4 in objective function)" by the coefficient of x_1 in objective function.

Step 8: Key element (or pivot element). The element $\boxed{1}$ at the intersection of the incoming vector (that is, here column Y_1) and the row in which the minimum ratio falls (i.e. here row which begins with Y_4) is the key element. The next step is to make each element (other than the key element) of the key column (i.e. the column in which key element occurs. That is, here column of Y_1) zero by suitable operations. This can be done by subtracting 2nd row from first row.

In this way the 2nd simplex table is formed.

TABLE 2

B	C_B	X_B	Y_1 $(=\beta_2)$	Y_2 $(=\alpha_2)$	Y_3 $(=\beta_1)$	Y_4	*Minimum ratio* $= \dfrac{X_B}{Y_2}$
Y_3	0 (=coeff. of x_3 in objective function,	$4 - 2 = 2$ $= x_3$	$1 - 1$	$= 0$ $\boxed{2}$		1	$\dfrac{2}{2} = 1$ -1 →
Y_1	3 (= coeff. of x_1 in objective function,	$2 = x_1$	1	-1	0	1	$\dfrac{2}{-1} = -2$
$Z = C_B X_B$	$= 0 \times 2 + 3 \times 2$ $= 6$	x_j / c_j / Δ_j	2 / 3 / 0	0 / 2 / 5	2 / 0 / 0	0 / 0 / -3	
				↑ Incoming vector	↓ outgoing vector		

$$\Delta_1 = c_1 - C_B\, Y_1 = 3 - (0, 3) \times (0, 1) = 3 - (0 \times 0 + 3 \times 1) = 0$$

$$\Delta_2 = c_2 - C_B\, Y_2 = 2 - (0, 3) \times (2, -1) = 5$$

$$\Delta_3 = c_3 - C_B\, Y_3 = 0 - (0, 3)\,(1, 0) = 0$$

$$\Delta_4 = c_4 - C_B\, Y_4 = 0 - (0, 3)\,(-1, 1) = -3$$

We see that Δ_2 (= 5) is + ve, so the solution is not optimal. Since $\Delta_2 = 5$ is the largest of Δ_j's, therefore Y_2 (= α) is the incoming vector, the minimum ratio is 2/2 (negative ratio not considered). So the outgoing vector is Y_3 (as the minimum ratio 2/2 occurs in row Y_3. In the next table Y_3 of column B will be replaced by y_2 and 0 (= coefficient of x_3 in objective function) of column C_B by corresponding value 2, the coefficient of x_2 in objective function.

The key element is $\boxed{2}$ at the intersection of column of incoming vector and row of outgoing vector.

Now make each element (other than key element) of key column zero. This can be done as below

Here,

	X_B	Y_1	Y_2	Y_3	Y_4
Y_3	2	0	$\boxed{2}$	1	–1
Y_4	2	1	–1	0	1

Divide first row by 2 inorder to make key element 1

	X_B	Y_1	Y_2	Y_3	Y_4
Y_3	1	0	1	$\frac{1}{2}$	$-\frac{1}{2}$
Y_4	2	1	–1	0	1

Add first row to 2nd row

	X_B	Y_1	Y_2	Y_3	Y_4
Y_3	1	0	1	$\frac{1}{2}$	$-\frac{1}{2}$
Y_4	3	1	0	$\frac{1}{2}$	1

Now, the 3rd simplex table in which β_1 (= Y_3) is replaced by a_2 (= Y_2) is as follows:

TABLE 3

B	C_B	X_B	Y_1 (= β_2)	Y_2 (= β_1)	Y_3	Y_4	*Minimum ratio*
Y_2	2 (= Coeff. x_2 in obj. fn.)	$1 = x_2$	0	1	$\frac{1}{2}$	$-\frac{1}{2}$	
Y_1	3 (= Coeff. of x_1 in obj. fn.)	$3 = x_1$	1	0	$\frac{1}{2}$	$-\frac{1}{2}$	
$Z = C_B X_B$		x_j	3	1	0	0	
$= 2 \times 1 + 3 \times 3$		c_j	3	2	0	0	
$= 11$		Δ_j	0	0	–5/2	$-\frac{1}{2}$	

Δ_j are calculated as above

$$\Delta_1 = c_1 - C_B Y_1 = 3 - (2, 3)(0, 1) = 0$$

$$\Delta_2 = c_2 - C_B Y_2 = 2 - (2, 3)(1, 0)$$

$$= 2 - 2 = 0$$

$$\Delta_3 = c_3 - C_B Y_3 = 0 - (2, 3)\left(\frac{1}{2}, \frac{1}{2}\right)$$

$$= 0 - \left(2 \times \frac{1}{2} + 3 \times \frac{1}{2}\right) = -\frac{5}{2}$$

$$\Delta_4 = c_4 - C_B Y_4 = 0 - (2, 3) \times \left(-\frac{1}{2}, \frac{1}{2}\right)$$

$$= 0 - \left[2 \times \left(-\frac{1}{2}\right) + 3 \times \frac{1}{2}\right] = -\frac{1}{2}$$

Since each Δ_j is either zero or less than 0, therefore this solution is optimal.

The optimal solution is given by

$$x_1 = 3, x_2 = 1 \text{ and}$$

$$\text{Max. } Z = 3x_1 + 2x_2$$

$$= 3 \times 3 + 2 \times 1 = 9 + 2 = 11$$

2. Use simplex method to solve the following L.P. problem.

Max. $Z = 7x_1 + 5x_2$

Subject to $x_1 + 2x_2 \leq 6$

$$4x_1 + 3x_2 \leq 12$$

$$x_1, x_2 \geq 0$$

[R.U. B.Sc. '87; V.B.U. Part II Hons. '99]

Solution:

Step 1: The given L.P.P. is problem of maximization.

Step 2: All the b_1's are already positive.

Step 3: The inequalities are converted into equations by introducing slack variables x_3 and x_4 as follows

$$x_1 + 2x_2 + x_3 = 6$$

$$4x_1 + 3x_2 + x_4 = 12$$

The new objective function is

$$Z = 7x_1 + 5x_2 + 0x_3 + 0x_4$$

Here, $c_1 = 7, c_2 = 5, c_3 = 0, c_4 = 0$

Equations (1) can be put as

$$Ax = b \text{ where}$$

$$A = \begin{bmatrix} 1 & 2 & 1 & 0 \\ 4 & 3 & 0 & 1 \end{bmatrix},$$

$$x = \begin{bmatrix} x_1 \\ x_2 \\ x_3 \\ x_4 \end{bmatrix},$$

$$b = \begin{bmatrix} 6 \\ 12 \end{bmatrix}$$

or, $[a_1, a_2, a_3, a_4]\, x = b$

$$a_1 = \begin{bmatrix} 1 \\ 4 \end{bmatrix}, a_2 = \begin{bmatrix} 2 \\ 3 \end{bmatrix}, a_3 = \begin{bmatrix} 1 \\ 0 \end{bmatrix}$$

where

$$a_4 = \begin{bmatrix} 0 \\ 1 \end{bmatrix}$$

Step. 4: Taking $x_1 = 0$, $x_2 = 0$, we get $x_3 = 6$, $x_4 = 12$ as starting B.F.S.

Step 5: Construct the starting simplex table

TABLE 1

B	C_B	X_B	Y_1 $(=a_1)$	Y_2 $(=a_2)$	Y_3 $(=\beta_1)$	Y_4 $(=\beta_2)$	*Minimum Ratio* $= \frac{X_B}{Y_1}$
Y_3	0	6	1	2	1	0	$\frac{6}{1} = 6$
Y_4	0	12	$\boxed{4}$	3	0	1	$\frac{12}{4} = 3 \rightarrow$
$Z = C_B X_B$		x_j	0	0	6	12	
$= 0 \times 6 + 0 \times 12$		c	7	5	0	0	
$= 0$		Δ_j	7	5	0	0	
			↑ Incoming vcctor			↓ outgoing vector	

Step 6:

$$\Delta_1 = c_1 - C_B Y_1 = 7 - (0, 0)(1, 4) = 7$$
$$\Delta_2 = c_2 - C_B Y_2 = 5 - (0, 0)(2, 3) = 5$$
$$\Delta_3 = c_3 - C_B Y_3 = 0 - (0, 0)(1, 0) = 0$$
$$\Delta_4 = c_4 - C_B Y_4 - 0 - (0, 0)(0, 1) = 0$$

Step 7: Since all Δ's are not less than or equal to zero, therefore $x_1 = 0$, $x_2 = 0$, $x_3 = 6$, $x_4 = 12$ do not give optimal solution. Clearly, α_1 is incoming vector and $\beta_2 = (= Y_4)$ is outgoing vector. The key element is $\boxed{4}$. Divide the key row by 4 in order to make the key element 1 and then subtract it from the first row (i.e. row of Y_3) to make all element (other than key element) of key column zero. In other words perform operation $R_1 - \frac{1}{4} R_2$.

Then the 2nd simplex table is as below.

TABLE 2

B	C_B	X_B	Y_1	Y_2	Y_3	Y_4
Y_3	0	3	0	$\frac{5}{4}$	1	$-\frac{1}{4}$
Y_1	7	3	1	$\frac{3}{4}$	0	$\frac{1}{4}$
$Z = C_B X_B$ $= 0 \times 3 + 7 \times 3$ $= 21$		x_j	3	0	3	0
		c_j	7	5	0	0
		Δ_j	0	$-\frac{1}{4}$	0	$-\frac{7}{4}$

Here, $\Delta_1 = c_1 - C_B Y_1 = 7 - (0, 7) \times (0, 1) = 7 - 7 = 0$

$$\Delta_2 = c_2 - C_B Y_2 = 5 - (0, 7)\left(\frac{5}{4}, \frac{3}{4}\right)$$

$$= 5 - \left(0 \times \frac{5}{4} + 7 \times \frac{3}{4}\right) = 5 - \frac{21}{4} = \frac{-1}{4}$$

$$\Delta_3 = C_3 - C_B Y_3 = 0 - (0, 7)(1, 0) = 0$$

$$\Delta_4 = C_4 - C_B Y_4 = 0 - (0, 7)\left(\frac{-1}{4}, 0\right) = 0$$

We see that each Δ is either zero or less than zero. Hence,

$x_1 = 3$, $x_2 = 0$, give maximum value of z.

$$Z \text{ max.} = 7x_1 + 5x_2$$

$$= 7 \times 3 + 5 \times 0$$

$$= 21.$$

3. Solve by Simplex method

Maximize $Z = 10x + 15y + 20z$

Subject to $10.7x + 5y + 2z \leq 2705$

$5.4x + 10y + 4z \leq 2210$

$0.7x + 1y \times 2z \leq 445$

and $x, y, z \geq 0$

Solution:

First convert the inequalities into equations by adding the slack variables S_1, S_2, S_3 then the problem becomes

Max. $Z = 10x + 15y + 20z + 0S_1 + 0S_2 + 0S_3$

Subject to
$$10.7x + 5y + 2z + 1S_1 + 0S_2 + 0S_3 = 2705$$
$$5.4x + 10y + 4z + 0S_1 + 1S_2 + 0S_3 = 2210$$
$$0.7x + 1y + 2z + 0S_1 + 0S_2 + 1S_3 = 445$$

and $x, y, z, S_1, S_2, S_3 \geq 0$

Starting Simplex Table

TABLE 1

(*i*) B	(*ii*) C_B	(*iii*) X_B	10 (*iv*) X	15 (*v*) Y	20 (*vi*) Z	(*vii*) S_1	(*viii*) S_2	(*ix*) S_3
Program (Slack variables introduced)	Profit per unit (co-eff. of slack variables in objective function)	Quantity Magnitude of slack variables by taking $x = y = z = 0$	co-eff. of x	co-eff. of y	co-eff. of z			
S_1	0	2,705	10.7	5	2	1	0	0
S_2	0	2,210	5.4	10	4	0	1	0
S_3	0	445	0.7	1	2	0	0	1

Explanation:

(i) Under the column programme; put the slack variables S_1, S_2, S_3.

(ii) Under column "Profit per unit", put the coefficients of S_1, S_2, S_3 in the objective function z.

(iii) Under the column "quantity" write down the values of S_1, S_2, S_3 when $x = y = z = 0$ are substituted in the equations. The first equation is

$$10.7x + 5y + 2z + 1S_1 + 0S_2 + \ 2,705$$

$\therefore$ $S_1 = 2,750$

In the same way $S_2 = 2,210$

$S_3 = 445$

(iv) Under the column X, put the co-eff. of x in the equation. Then co-eff. of x are 10.7, 5.4, 0.7. Similarly co-eff. of y, z, S_1, S_2, S_3 of the equations have been written under the column Y, Z, S_1, S_2, S_3.

(v) Write down the co-eff. of x, y, z, S_1, S_2, S_3 in the objective function Z at the top of columns.

(vi) In this first programme the total profit contribution is zero

[= 0(2,705) + 0(2, 210) + 0(4.45)]

Second simplex table

TABLE 2

B Programme	C_B Profit per unit	X_B Quantity	10 X	15 Y	20 Z	0 S_1	0 S_2	0 S_3	Smallest Ratio
S_1	0	2,705	0.7	5	2	1	0	0	1352.5
S_2	0	2,210	5.4	10	4	0	1	0	552.5
S_3	0	445	0.7	1	[2]	0	0	1	222.5 →
Net evaluation row			10	15	20 ↑	0	0	0 ↓	

(basic row or opportunity cost)

Explanation:

(i) Calculation of net evaluation row under the column X.
= (Co-eff. of X in objective function; i.e. numbers at the top of X)— (product of entries under column "Profit per unit" and X)
$= 10 - (0 \times 10.7 + 0 \times 5.4 + 0 \times 0.7) = 10.$
Similarly, other entries in net evaluation row can be calculated.

(ii) Testing the optimality—For maximum value of Z, each and every entry in Net-evaluation row must be less than or equal to zero.

(iii) Key column—Locate the largest +ve number in the Net Evaluation Row. The column under which this number lies is called key column here, 20 is largest in +ve in the Base row. So the column under Z is key column.

(iv) Smallest Ratio—Enteries under the column are obtained by dividing the enteries under the column "Quantity" by the corresponding non-negative entries of the key column. This can be shown as below—

$$\text{For row } S_1 \quad \frac{2705}{2} = 1352.6$$

$$\text{For row } S_2 \quad \frac{2210}{4} = 552.5$$

$$\text{For row } S_3 \quad \frac{445}{2} = 222.5$$

The row in which the smallest ratio (222.5) lies is called key row.

(v) Key Number—The number at the intersection of key column and key row is called the key number. Here 2 is the key number.

Third Simplex Table

TABLE 3

B *Programme*	C_B *Profit per unit*	X_B *Quantity*	*X*	*Y*	*Z*	S_1	S_2	S_3	*Smallest Ratio*
S_1	0	2260	10	4	0	1	0	– 1	$\frac{2260}{4} = 656$
S_2	0	1330	4	[8]	0	0	1	– 2	$\frac{1320}{8} = 165$ →
Z	20	222.5	.35	.5	1	0	0	.5	$\frac{222 \cdot 5}{\cdot 5} = 445$
Net evaluation row			3	5 ↑ incoming	0	0	0 ↓ outgoing	–10	

Explanation

(i) Since the key column in Table 2 is under Z, replace S_2 in "Programme" column by Z and put the value of $Z(= 20)$ in "Profit per unit" column.

(ii) Transformation of key row—Divide all the numbers in the key row by the key number. Divide the numbers in row S_3 of Table 2 by the key number 2 and put it exactly in the same position in Table 3. The number in row Z of Table 3 are

$$222.5 \left(=\frac{445}{2}\right), .35 \left(=\frac{7}{2}\right), .5 \left(\frac{1}{2}\right)$$

$$1 \left(=\frac{2}{2}\right), 0 \left(=\frac{0}{2}\right), 0 \left(=\frac{0}{2}\right); 5 \left(=\frac{1}{2}\right)$$

(iii) Transformation of non-key rows—In Table 2 the row S_1, and row S_2 are non-key rows. The transformation of these numbers can be made by the following rule

New row number = old row number (Corresponding number in key row × corresponding fixed ratio) where fixed ratio= old row number in key column/key number.

The new row S_1 for Table 3 is derived as follows (Corresponding fixed ratio = $\frac{2}{2} = 1$).

Old row number	– Corresponding no. in key row	× Corresponding fixed ratio	= New row number
2705	– 445	× 1	= 2260
10.7	– 0.7	× 1	= 10
5	– 1	× 1	= 4
2	– 2	× 1	= 0
1	– 0	× 1	= 1
0	– 0	× 1	= 0
0	– 1	× 1	= – 1

Thus 2260, 10, 4, 0, 1, 0, – 1 are shown as numbers in row S_1 in table 3. Similarly, row S_2 of Table 3 can be obtained for which the fixed ratio is 2(= 4/2).

(iv) The net evaluation row

5 is the largest +ve number in net evaluation row. So the key column is the column under *Y*. The smallest number under the column "smallest ratio" is 165 $\left(=\frac{1320}{8}\right)$ So the key row (outgoing) is the row S_2. The key number is 8 which is the intersection of key row and key column.

As the net evaluation row of Table 3 shows, we still have two separate +ve numbers (3, 5). Hence programme *Z* contained in Table 3 is not an optimum programme. This calls for constructing a new Programme and therefore, deriving a new Table 4. The procedure of constructing Table 4 is exactly the same as was followed in Table 3.

FOURTH SIMPLEX TABLE

TABLE 4

B Programme	C_B Profit per unit	X_B Quantity	10 *X*	15 *Y*	20 *Z*	0 S_1	0 S_2	0 S_3	Smallest ratio
S_1	0	1600	[8]	0	0	1	– 0.5	0	$\frac{1600}{8} = 200$ →
Y	15	165	0.5	1	0	0	.125	– .25	$\frac{165}{.5} = 330$
Z	20	140	0.1	0	1	0	– 0.0620	0.65	$\frac{140}{.1} = 1400$
Net evaluation row			.5 ↑ Incoming	0	0	0 ↓ outgoing	–.635	– 8.75	

Explanation:

(1) The second row of Table 4 i.e. row Y is derived from the 2nd row of Table 3 (row S_2) by dividing all in row S_2 by the key number 8.

Row Y of Table 4 becomes 165, .5, 1, 0, 0, .125, –.25.

(2) The non-key row of Table 3 are transformed as follows

Old row number – corresponding number × corresponding = new row

		in key row		fixed ratio		number
2260	—	1320	×	.5	=	1600
10	—	4	×	.5	=	8
4	—	8	×	.5	=	0
0	—	0	×	.5	=	0
1	—	0	×	.5	=	1
0	—	1	×	.5	=	–.5
–1	—	(–2)	×	.5)	=	5

The row S_1 of Table 4 becomes 1600, 8, 0, 0, 1, –.5, 0

Calculations for row Z $\left(\text{fixed ratio} = \frac{.5}{8} = .0625\right)$

Old row number – Corresponding number × Corresponding = New row

		in key row		fixed ratio		number
222.5	—	1320	×	.0625	=	40
.35	—	4	×	.0625	=	.1
.5	—	8	×	.0625	=	0
1	—	0	×	.0625	=	1
0	—	0	×	.0625	=	0
0	—	1	×	.0625	=	– 0.062
.5	—	–2	×	.0625	=	.625

The row Z becomes

140.1	0	1	0	– 0.062	.625

(3) The Net evaluations row

10	—	0 × 8	—	15 × .5	—	20. × .1	=	.5
15	—	0 × 0	—	15 × 1	—	20 × 0	=	0
20	—	0 × 0	—	15 × 0	—	20 × 1	=	0
0	—	0 × 1	—	1 × 0	—	20 × 0	=	0
0	—	0 × (–.5)	—	15 × .125	—	20 × (–.062)	=	–.635
0	—	0 × 0	—	15 × (–.25)	—	20 × (–.625)	=	– 8.75

We see that the Net evaluations row in Table 4 still containing one +ve number (.5). Hence this programme does not give the optimum programme. Column × Column is key column (incoming variable) and row S_1 is key row (incoming variable), and 8 is key number. Then we form a new Table 5.

Table 5. A before the derivation of the programme Table 4 is calculated the key row of table 4 (row S_1) is transformed in row X.

Fifth Simplex Table

TABLE 5

B *Progr-amme*	C_B *Profit per unit*	X_B *Quantity*	10 *X*	15 *Y*	20 *Z*	0 S_1	0 S_2	0 S_3
X	10	200	1	0	0	.125	–.062	0
Y	15	65	0	1	0	–.062	.156	–.25
Z	20	120	0	0	1	–.012	–.056	.625
Net evaluation row			0	0	0	– 0.625	–.523	– 8.75

Explanation of Table 5

We see that the enteries in the net evaluation row are ≤ 0. Hence this is optimal programme.

The maximum Z is $10 \times 200 + 15 \times 65 + 20 \times 120 = 5375$

Example 3. Solve by Simplex method the L.P.P.

$$\text{Max.} \quad Z = 3x_1 + 5x_2 + 4x_3$$

Subject to

$$2x_1 + 3x_2 \leq 8$$
$$2x_2 + 5x_3 \leq 10$$
$$3x_1 + 2x_2 + 4x_3 \leq 15$$
$$x_1, x_2, x_3 \geq 0$$

[R.U. Condensed course Hons. 80, Meerut 83, 84, 88, L.P. 73 (S), 86, 69, 71, 74S, 75S
M. Com. 83 (P), Delhi 64, Shivajee 76, 77, [SI 70 Raj T.D.C. 77 Kuru 83 Agra 80, 84 I.C.
W.A. June 80 Summer C.A. Nov. 77, Kanpur 70, Meerut Stat 76]

Solution: By adding the slack variables S_1, S_2, S_3 the problem becomes

$$\text{Max. } Z = 3x_1 + 5x_2 + 4x_3 + 0S_1 + 0S_2 + 0S_3$$

Subject to $\quad 2x_1 + 3x_2 + S_1 + 0S_2 + 0S_3 = 8$

$$2x_2 + 5x_3 + 0S_1 + S_2 + 0S_3 = 10$$

$$3x_1 + 2x_2 + 4x_3 + 0S_1 + 0S_2 + S_3 = 15$$

Taking $x_1 = 0, x_2 = 0, x_3 = 0$, we get from the three equations $S_1 = 8, S_2 = 10, S_3 = 15$ which is starting B.F.S. Following the procedure of Simplex, we get the

Starting simplex Table

TABLE 1

B Pro-gramme	*CB Profit per unit*	*X_B Quantity*	*3* x_1	*5* x_2	*4* x_3	*0* S_1	*0* S_2	*0* S_3	*Smallest ratio*
S_1	0	8	2	[3]	0	1	0	0	$\frac{8}{3}$ →
S_2	0	10	0	2	5	0	1	0	$\frac{10}{2} = 5$
S_3	0	15	3	2	4	0	0	1	$\frac{15}{2} = 7.5$
Net evaluation row			3	5 ↑ Key column	4	0	0	0	

Second Simplex Table

TABLE 2

B Pro-gramme	*C_B Profit per unit*	*C_B Quantity*	3 x_1	5 x_2	4 x_3	0 S_1	0 S_2	0 S_3	*Smallest ratio*
S_1	5	$\frac{8}{3}$	$\frac{2}{3}$	1	0	$\frac{1}{3}$	0	0	$\frac{8/3}{0} = \infty$
S_2	0	$\frac{14}{3}$	$-\frac{4}{3}$	0	5	$-\frac{2}{3}$	1	0	$\frac{14/3}{5} = \frac{14}{15}$ →
S_3	0	$\frac{29}{3}$	$\frac{5}{3}$	0	4	$-\frac{2}{3}$	0	1	$\frac{29}{12}$
Net evaluation row			$\frac{-1}{3}$	0	4 ↑ Incoming	$\frac{-5}{3}$	0 ↓ Outgoing	0	

Third Simplex Table

TABLE 3

Program	Profit per unit	Quantity	3 x_1	5 x_2	4 x_3	0 S_1	0 S_2	0 S_3	Smallest ratio
x_2	5	$\frac{8}{3}$	$\frac{2}{3}$	1	0	$\frac{1}{3}$	0	0	$\frac{8}{3} \times \frac{3}{2} = 4$
x_3	4	$\frac{14}{15}$	$-\frac{4}{15}$	0	1	$-\frac{2}{15}$	$\frac{1}{5}$	0	$-\frac{7}{2}$
S_3	0	$\frac{89}{15}$	$\boxed{\frac{41}{15}}$	0	0	$-\frac{2}{15}$	$-\frac{4}{5}$	1	$\frac{89}{41} \rightarrow$
Net evaluation row		$\frac{11}{15}$	0	0	$\frac{-17}{15}$	$-\frac{4}{5}$	0		

↑ Incoming ↓ Outgoing

Fourth Simplex Table

TABLE 4

B Programme	C_B Profit per unit	X_B Quantity	3 x_1	5 x_2	4 x_3	0 S_1	0 S_2	0 S_3
x_2	5	$\frac{50}{41}$	0	1	0	$\frac{15}{4}$	$\frac{8}{41}$	$-\frac{10}{41}$
x_3	4	$\frac{62}{41}$	0	0	1	$\frac{6}{41}$	$\frac{5}{41}$	$\frac{4}{41}$
x_1	3	$\frac{89}{41}$	1	0	0	$-\frac{2}{41}$	$-\frac{12}{41}$	$\frac{15}{41}$
Net Evaluation row			0	0	0	$\frac{-45}{41}$	$\frac{-24}{41}$	$\frac{-11}{41}$

Since all the numbers in net evaluation row ≤ 0

Hence this programme gives optimal solution

$$\therefore \quad \text{Max. } Z = 3x_1 + 5x_2 + 4x_3$$

$$= 3 \times \frac{89}{41} + 5 \times \frac{50}{41} + 4 \times \frac{62}{41}$$

$$= \frac{765}{41}$$

for $x_1 = \frac{89}{41}$

$x_2 = \frac{50}{41}$

$x_3 = \frac{62}{41}$ Ans.

Example 4. Solve

$$\text{Max. } Z = 5x_1 + 3x_2$$
$$3x_1 + 5x_2 \leq 15$$
$$5x_1 + 2x_2 \leq 10$$
$$\text{and} \quad x_1, x_2 \geq 0$$

(Raj. M.Sc. 73, T.D.C. 76, Delhi 69)
(V B U Part II Hons. 98)

Solution: This is problem of maximization.

Now, the equations obtained by introducing slack variables x_3, x_4 are as follows.

$$3x_1 + 5x_2 + x_3 = 15$$
$$5x_1 + 2x_2 + x_4 = 10$$

Taking $x_1 = 0, x_2 = 0$ we get $x_3 = 15$ and $x_4 = 10$ which is starting B.F.S. Adopting the method of simplex, we get the

Starting simplex Table

TABLE 1

B	C_B	X_B	Y_1 $= \alpha$	Y_2 $= \alpha_1$	Y_3 $= \beta_1$	Y_4 $= \beta_2$	*Mini. ratio* $= \frac{X_B}{Y_1}$
Y_3	0	15	3	5	1	0	$\frac{15}{3} = 5$
Y_4	0	10	5	2	0	1	$\frac{10}{5} = 2$ (mini.) ⟶
		x_j	0	0	15	10	
		c_j	5	3	0	0	
		Δ_j	5	4	0	0	
			↑ incoming vector			↓ outgoing vector	

$\Delta_1 = C_1 - C_B Y_1 = 5 - (0, 0)(3, 5) = 5$ $\Delta_3 = C_3 - C_B Y_3 = 0 - (0, 0)(0, 0) = 0$

$\Delta_2 = C_2 - C_B Y_2 = 4 - (0, 0)(5, 2) = 4$ $\Delta_4 = C_4 - C_B Y_4 = 0 - (0, 0)(0, 1) = 0$.

Since all Δj's are not less or equal to zero, therefore, the solution is not optimal.

Since, $\Delta_1 = 5$ is max. of Δ_1, Δ_2

$\therefore$ $\alpha_1 (= Y_1)$ is incoming vector, and by mini ratio rule, $\beta_2 (= Y_4)$ is out going vector.

$\therefore$ Key element is $Y_{21} (= \alpha_{21})$ which is equal to 5 (not equal to 1).

$\therefore$ In order to bring β_2 in place of α_1 . the intermediate tables are as follows

	X_B	Y_1	Y_2	Y_3	Y_4
Y_3	15	3	5	1	0
Y_4	10	$\boxed{5}$	2	0	1

	X_B	Y_1	Y_2	Y_3	Y_4
Y_2	9	0	$\frac{19}{5}$	1	$\frac{-3}{5}$
Y_4	2	1	$\frac{2}{5}$	0	$\frac{1}{5}$

Now, we construct 2nd simplex table.

TABLE 2

B	C_B	X_B	Y_1	Y_2	Y_3	Y_4	Mini, X_B ratio = Y_3
Y_3	0	9	0	$\frac{19}{5}$	1	$-\frac{3}{5}$	$\frac{9 \times 5}{19} = \frac{45}{19}$ →
Y_1	5	2	1	$\frac{2}{5}$	0	$\frac{1}{5}$	$\frac{2 \times 5}{2} = 5$
$Z = C_B X_B = 10$		x_j	2	0	9	0	
		c_j	5	3	0	0	
		Δj	0	1	0	−1	
			↑ incoming vector		↓ outgoing vector		

$$\Delta_2 = C_2 - C_B\ Y_2 = 3 - (0, 5)\left(\frac{19}{5}, \frac{2}{5}\right) = 1$$

$$\Delta_4 = C_4 - C_B\ Y_4 = 0 - (0, 5)\left(-\frac{3}{5}, \frac{1}{5}\right) = -1$$

Since all Δj are not less than or equal to zero, therefore, this solution is not optimal.

Since, $\Delta_2 = 1$ is max. of the Δj's

$\therefore$ a_2 (= Y_2) is the incoming vector and by the minimum ratio rule we find β_1 = (= Y_3) is outgoing vector.

In order to bring Y_2 in place of β_1 (= Y_3) we make the following intermediate table.

	X_B	Y_1	Y_2	Y_3	Y_4
Y_3	9	0	$\frac{19}{5}$	1	$-\frac{3}{5}$
Y_1	1	2	$\frac{2}{5}$	1	$-\frac{1}{5}$

	X_B	Y_1	Y_2	Y_3	Y_4
Y_3	$\frac{45}{19}$	0	1	$\frac{5}{19}$	$-\frac{3}{19}$
Y_1	$\frac{20}{19}$	1	0	$-\frac{2}{19}$	$\frac{5}{19}$

The third Simplex table in which β_1 (= Y_3) is replaced by Y_2 as follows

TABLE 3

B	C_B	X_B	Y_1	Y_2	Y_3	Y_4	*Mini. ratio*
Y_2	3	$\frac{45}{19}$	0	1	$\frac{5}{19}$	$-\frac{3}{19}$	
Y_1		$\frac{20}{19}$	1	0	$-\frac{2}{19}$	$\frac{5}{19}$	
		x_j	$\frac{20}{19}$	$\frac{45}{19}$	0	0	
$Z = C_B X_B$		c_j	5	3	0	0	
$= \frac{235}{19}$		Δj	0	0	$\frac{-5}{19}$	$-\frac{16}{19}$	

$$\Delta_3 = C_3 - C_B Y_3 = 0 - (3, 5)\left(\frac{5}{19}, -\frac{2}{19}\right)$$

$$= -\frac{15}{19} + \frac{10}{19} = \frac{-15+10}{19} = -\frac{5}{19}$$

$$\Delta_4 = C_4 - C_B Y_4 = 0 - (3, 5)\left(-\frac{3}{19}, \frac{5}{19}\right)$$

$$= 0 + \frac{9}{19} - \frac{25}{19} = \frac{9-25}{19} = -\frac{16}{19}$$

Since all Δ_j for all non-zero variables are negative so this solution is optimal.

$$x_1 = \frac{20}{19}, \qquad x_2 = \frac{45}{19}$$

$$\therefore \qquad \text{Max. } z = \frac{235}{19} \qquad \text{Ans.}$$

Example 5. Solve the following L.P.P. by0 Simplex method

Minimize $z = x_1 - 3x_2 + 2x_3$

Subject to $3x_1 - x_2 + 2x_2]\leq 7$

$- 2x_1 + 4x_2 \leq 12$

$- 4x_1 + 3x_2 + 8x_3 \leq 10$

$x_1, x_2, x_3 \geq 0$

[Delhi 76, Roorkee M.Sc. (Elect) 77, T.D.C. 77 (Nov.) Meerut O.R. 70, 76, 87, Dibrugarh M.Sc. (Stat 76), Raj M.Sc. L.P., 85 Rohilkhand 84, Meerut C.P. 89, M.Sc. (Stat) 74]

Solution: First we convert the problem of minimization to maximization problem by taking the objective function as

$$z' = -z$$

Max. $z' = -z = -x_1 + 3x_2 - 2x_3$

Now, the equations obtained by introducing slack variables x_4, x_5, x_6 are as follows

$$3x_1 - x_2 + 2x_3 + x_4 = 7$$
$$- 2x_1 - 4x_2 + x_5 = 12$$
$$- 4x_1 + 3x_2 + 8x_3 + x_6 = 10$$

Taking $x_1 = 0, x_2 = 0, x_3 = 0$, we get

$x_4 = 7, x_5 = 12, x_6 = 10$ which is starting B.F.S.

Employing the method of simplex, we get the following table.

Starting Simplex Table

TABLE 1

B	C_B	X_B	$Y_1 = \alpha_1$	$Y_2 = \alpha_2$	$Y_3 = \alpha_3$	$Y_4 = \beta_1$	$Y_5 = \beta_2$	$Y_6 = \beta_3$	*Mini. ratio* $= (X_B/Y_2)$
Y_4	0	2	3	$\boxed{-1}$	2	1	0	0	$\frac{7}{-1}$ (Neg)
Y_5	0	12	−2	4	0	0	1	0	3 (mini.) →
Y_6	0	10	−4	3	8	0	0	1	$\frac{10}{3}$
$Z = C_B X_B$		x_j	0	0	0	7	12	10	
		c_j	−1	3	−2	0	0	0	
$= 0$		Δ_j	−1	3	−2	0	0	0	

↑ incoming vector (under Y_2) ↓ outgoing vector (under Y_5)

Second Simplex Table

TABLE 2

B	C_B	X_B	Y_1	Y_2 $(=\beta_2)$	Y_3	Y_4 $(=\beta_1)$	Y_5	Y_6 $(=\beta_3)$	mini, X_B ratio $=\frac{X_B}{Y_1}$
Y_4	0	10	$\frac{5}{2}$	0	2	1	$\frac{1}{4}$	0	4 (mini) →
Y_2	3	3	$-\frac{1}{2}$	1	0	0	$\frac{1}{4}$	0	−6
Y_6	0	1	$-\frac{5}{2}$	0	8	0	$-\frac{3}{4}$	1	$-\frac{2}{5}$
$Z' = C_B X_B$		x_j	0	3	0	10	0	1	
		c_j	−1	3	−2	0	0	0	
	= 9	Δ_j	$\frac{1}{2}$	0	−2	0	$-\frac{2}{4}$	0	

↑ incoming vector ↓ outgoing vector

Third Simplex Table.

TABLE 3

B	C_B	X_B	Y_1 $(=\beta_1)$	Y_2 $(=\beta_2)$	Y_3	Y_4	Y_5	Y_6 $(=\beta_3)$	mini. ratio
Y_1	−1	4	1	0	$\frac{4}{5}$	$\frac{2}{5}$	$\frac{1}{10}$	0	
Y_2	3	5	0	1	$\frac{2}{5}$	$\frac{1}{5}$	$\frac{3}{10}$	0	
Y_6	0	11	0	0	10	1	$-\frac{1}{2}$	1	
$Z' = C_B X_B$		x_j	4	5	0	0	0	11	
= 11		c_j	−1	3	−2	0	0	0	
		Δ_j	0	0	$-\frac{12}{5}$	$-\frac{1}{5}$	$-\frac{4}{5}$	0	

Since all Δ_j for all non-zero variables are negative, so this solution is optimal.

Optimal solution is $x_1 = 4, x_2 = 5, x_2 = 0$

and mini. $Z = -Z = -11.$ Ans.

Example 6. Solve

$$\begin{aligned} \text{Max.} \quad & Z = 2x_1 + 5x_2 + 7x_3 \\ \text{Subject} \quad & 3x_1 + 2x_2 + 4x_2 \le 100 \\ & x_1 + 4x_2 + 2x_3 \le 100 \\ & x_1 + x_2 + 3x_3 \le 100 \\ & x_1, x_2, x_3 \ge 0 \end{aligned}$$

[V.B.U. Part III Hond. 97 (Old Spl.), Meerut 73 S, 74, Agra 86],

Solution: The equations obtained by introducing slack variables x_4, x_5, x_6 are as follows

$$3x_1 + 2x_2 + 4x_3 + x_4 = 100$$
$$x_1 + 4x_2 + 2x_3 + x_5 = 100$$
$$x_1 + x_2 + 3x_3 + x_6 = 100$$

Taking $x_1 = 0, x_2 = 0$

$x_3 = 0$, we get

$$x_4 = 100, x_5 = 100, x_6 = 100$$

Which is starting B.F.S. The simplex procedure gives the following.

Starting Simplex Table

TABLE 1

B	C_B	X_B	Y_1 $(=\alpha_1)$	Y_2 $(=\alpha_2)$	Y_3 $(=\alpha_3)$	Y_4 $(=\beta_1)$	Y_5 $(=\beta_2)$	Y_6 $(=\beta_3)$	*Mini. ratio*
									(mini.)
Y_4	0	100	3	2	$\boxed{4}$	1	0	0	25 →
Y_5	0	100	1	4	2	0	1	0	50
Y_6	0	100	1	1	3	0	0	1	$\frac{100}{3}$
$Z = C_B X_B$		x_j	0	0	0	100	100	100	
		c_j	2	5	7	0	0	0	
$= 0$		Δ_j	2	5	7	0	0	0	
					↑ incoming vector	↓ outgoing vector			

Second Simplex Table

TABLE 2

B	C_B	X_B	Y_1	Y_2	Y_3 $(=\beta_1)$	Y_4	Y_5 $(=\beta_2)$	Y_6 $(=\beta_3)$	*Mini ratio*
Y_3	7	25	$\frac{3}{4}$	$\frac{1}{2}$	1	$\frac{1}{4}$	0	0	50
Y_5	0	50	$-\frac{1}{2}$	$\boxed{3}$	0	$-\frac{1}{2}$	1	0	$\frac{50}{3}$ (mini.) →
Y_6	0	25	$-\frac{5}{4}$	$-\frac{1}{2}$	0	$-\frac{3}{4}$	0	1	(Neg)
$Z = C_B X_B$		x_j	0	0	25	0	50	25	
$= 175$		c_j	2	5	7	0	0	0	
		Δ_j	$-\frac{13}{4}$	$\frac{3}{2}$	0	$-\frac{7}{4}$	0	0	
				↑ incoming vector			↓ outgoing vector		

Third Simplex Table

TABLE 3

B	C_B	X_B	Y_1	Y_2 $(=\beta_2)$	Y_3 $(=\beta_1)$	Y_4	Y_5	Y_6	*Mini. ratio*
Y_3	7	$\frac{50}{3}$	$\frac{5}{6}$	0	1	$\frac{1}{3}$	$-\frac{1}{6}$	0	
Y_2	5	$\frac{50}{3}$	$-\frac{1}{6}$	1	0	1	$\frac{1}{3}$	0	
Y_6	0	$\frac{100}{3}$	$-\frac{4}{5}$	0	0	$-\frac{5}{6}$	$\frac{1}{6}$	1	
$Z = C_B X_B$		x_j	0	$\frac{50}{3}$	$\frac{50}{3}$	0	0	$\frac{100}{3}$	
$= 200$		c_j	2	5	7	0	0	2	
		Δ_j	-3	0	0	$-\frac{3}{2}$	$-\frac{1}{2}$	0	

Since all Δ_j's for zero variables are negative, so this solution is optimal.

$\therefore$ Optimal solution is $x_1 = 0,\quad x_2 = \frac{50}{3},\quad x_3 = \frac{50}{3}$

$\therefore$ Max. $z = 200$. Ans.

Example 7. Solved by simplex method

Max. $Z = 2x_1 + 4x_2$

Subjects to

$$2x_1 + 3x_2 \le 48$$
$$x_1 + 3x_2 \le 42$$
$$x_1 + x_2 \le 21$$
$$x_1, x_2 \ge 0$$

Solution: The equations obtained by introducing slack vari-ables x_3, x_4, x_5 are as follows

$$2x_1 + 3x_2 + x_3 = 48$$
$$x_1 + 3x_2 + x_4 = 42$$
$$x_1 + x_2 + x_5 = 21$$

Taking $x_1 = 0$, $x_2 = 0$ then $x_3 = 48, x_4 = 42, x_5 = 21$, which is the starting B.F.S.

Using the methods of simplex. We get the following table.

TABLE 1

B	C_B	X_B	Y_1 $(= \alpha_1)$	Y_2 $(= \alpha_2)$	Y_3 $(= \beta_1)$	Y_4 $(= \beta_2)$	Y_5 $(= \beta_3)$	*mini. ratio*
Y_3	0	48	2	3	1	0	0	$\frac{48}{3} = 16$
Y_4	0	42	1	$\boxed{3}$	0	1	0	$\frac{42}{3} = 14$ mini →
Y_5	0	21	1	1	0	0	1	$\frac{21}{1} = 21$
$z = C_B X_B$	x_j	0	0	48	42	21		
	c_j	2	4	0	0	0		
= 0	Δ_j	2	4	0	0	0		
			↑ incoming vector		↓ outgoing vector			

Second Simplex Table

TABLE 2

B	C_B	X_B	Y_1 $(\alpha = 1)$	Y_2 $(= \beta_2)$	Y_3 $(= \beta_1)$	Y_4	Y_5	*Mini. ratio*
Y_3	0	6	$\boxed{1}$	0	1	-1	0	$\frac{6}{1} = 6$ →
Y_2	4	14	$\frac{1}{3}$	1	0	$\frac{1}{2}$	0	$\frac{14}{\frac{1}{3}} = 42$
Y_5	0	7	$\frac{2}{3}$	0	0	$-\frac{1}{2}$	1	$\frac{21}{2}$
$Z = C_B X_B$	x_j	0	14	6	0	7		
	c_j	2	4	0	0	0		
= 56	Δ_j	$\frac{2}{3}$	0	0	$-\frac{4}{3}$	0		
			↓ incoming vector		↑ outgoing vector			

Third Simplex Table

TABLE 3

B	C_B	X_B	Y_1 $(=\beta_1)$	Y_2 $(=\beta_2)$	Y_3	Y_4	Y_5	*Mini-ratio*
Y_1	2	6	1	0	1	-1	0	
Y_2	4	12	0	1	$-\frac{1}{3}$	$\frac{2}{3}$	0	
Y_5	0	3	0	0	$-\frac{2}{3}$	$\frac{1}{3}$	1	
$Z = C_B X_B$		x_j	6	12	0	0	3	
		c_j	2	4	0	0	0	
$= 60$		Δ_j	0	0	$-\frac{2}{3}$	$-\frac{2}{5}$	0	

Thus all Δ_j's are ≤ 0. Hence, the solution is optimal.

$$Z \text{ max.} = 2 \times 6 + 4 \times 12$$

$$= 60$$

$\therefore$ for $x_1 = 6,$

$x_2 = 12.$

Example 8. Max. $z = 3x_1 + 5x_2 + 4x_3$

Subject to

$$2x_2 + 3x_3 \leq 18$$

$$2x_2 + 5x_3 \leq 18$$

$$3x_1 + 2x_2 + 4x_3 \leq 25$$

$$x_1, x_2, x_3 \geq 0$$

Solution: The equations obtained by introducing slack variables x_4, x_5, x_6 are as follows

$$0.x_1 + 2x_2 + 3x_3 + x_4 = 18$$

$$0.x_1 + 2x_2 + 5x_3 + x_5 = 18$$

$$3x_1 + 2x_2 + 4x_3 + x_6 = 25$$

Taking $x_1 = 0, x_2 = 0, x_3 = 0$ we get $x_4 = 18, x_5 = 18, x_6 = 25$ which is starting B.F.S. Following the procedure of simplex, we have the Starting Simplex Table.

TABLE 1

B	C_B	X_B	Y_1 $(=\alpha_1)$	Y_2 $(=\alpha_2)$	Y_3 $(=\alpha_3)$	Y_4 $(=\beta_1)$	Y_5 $(=\beta_2)$	Y_6 $(=\beta_3)$	*Mini ratio*
Y_4	0	18	0	2	3	1	0	0	$\frac{18}{2}=9$
Y_5	0	18	0	[2]	5	0	1	0	$\frac{18}{2}=9$ →
Y_6	0	25	3	2	4	0	0	1	$\frac{25}{2}=12.5$
$Z=C_B X_B$		x_j	0	0	0	18	18	25	
$=0$		c_j	3	5	4	0	0	0	
		Δ_j	3	5	4	0	0	0	
				↑ incoming vector			↓ outgoing vector		

Second Simplex Table

TABLE 2

B	C_B	X_B	Y_1 $(=\alpha_1)$	Y_2 $(=\beta_2)$	Y_3 $(=\alpha_3)$	Y_4 $(=\beta_1)$	Y_5	Y_6 $(=\beta_3)$	*Mini ratio*
Y_4	0	0	0	0	-2	1	$-\frac{1}{2}$	0	$\frac{0}{0}$
Y_2	5	9	0	1	$\frac{5}{2}$	0	$\frac{1}{2}$	0	$\frac{9}{0}$
Y_6	0	7	[3]	0	-1	0	-1	1	$\frac{7}{3}=2.3$ →
$Z=C_B X_B$		x_j	0	9	0	0	0	7	
		c_j	3	5	4	0	0	0	
$=45$		Δ_j	3	0	$-\frac{17}{2}$	0	$-\frac{5}{2}$	0	
			↑ Incoming vector				↓ Outdoing vector		

Third Simplex Table

TABLE 3

B	C_B	X_B	Y_1 $(=\beta_3)$	Y_2 $(=\beta_2)$	Y_3 $(=\alpha_3)$	Y_4 $(=\beta_1)$	Y_5	Y_6	Mini. ratio
Y_4	0	0	0	0	-2	1	-1	0	
Y_2	5	9	0	1	$\frac{5}{2}$	0	$\frac{1}{2}$	0	
Y_1	3	$\frac{7}{3}$	1	0	$-\frac{1}{3}$	0	$-\frac{1}{3}$	$\frac{1}{3}$	
$Z = C_B X_B$		x_j	$\frac{7}{3}$	9	0	0	0	0	
		c_j	3	5	4	0	0	0	
$= 52$		Δ_j	0	0	$-\frac{15}{2}$	0	$-\frac{3}{2}$	-1	

Since all Δ_j's are negative or zero so the solution is optimal.

$$x_1 = \frac{7}{3},\ x_2 = 9,\ x_3 = 0$$

$$\therefore \quad Z = 3x_1 + 5x_2 + 4x_3$$

$$= 3 \times \frac{7}{3} + 5 \times 9 + 4 \times 0$$

$$= 9 + 45 = 52 \quad \text{Ans.}$$

Example 9. Solve

$$\text{Max. } Z = 7x_1 + 5x_2$$

$$\text{Subject to} \quad x_1 + 2x_2 \le 6$$

$$4x_1 + 3x_2 \le 12$$

$$x_1, x_2 \ge 0$$

[R.U. B.Sc. Hons. '87; R.U.M.Sc. '82]

Solution: The equations obtained by using slack variables are as follows.

$$x_1 + 2x_2 + x_3 = 6$$

$$4x_1 + 3x_2 + x_4 = 12$$

Taking $x_1 = 0, x_2 = 0$ we get $x_3 = 6$ and $x_4 = 12$, which is starting B.F.S.

Starting Simplex Table

TABLE 1

B	C_B	X_B	Y_1 $(=\alpha_1)$	Y_2 $(=\alpha_2)$	Y_3 $(=\beta_1)$	Y_4 $(=\beta_2)$	Mini. ratio
Y_3	0	6	1	2	1	0	$\frac{6}{1}=6$
Y_4	0	12	$\boxed{4}$	3	0	1	$\frac{12}{4}=3$ (mini) $\rightarrow$
$Z = C_B X_B$		x_j	0	0	6	12	
		c_j	7	5	0	0	
$= 0$		Δ_j	7	5	0	0	
			↑ Incoming vector			↓ Out going vector	

Second Simplex Table

TABLE 2

B	C_B	X_B	Y_1 $(=\beta_2)$	Y_2 $(=\alpha_2)$	Y_3 $(=\beta_1)$	Y_4	Mini. ratio
Y_3	0	3	0	$\frac{5}{4}$	1	$-\frac{1}{4}$	
Y_1	7	3	1	$\frac{3}{4}$	0	$\frac{1}{4}$	
$Z = C_B X_B$		x_j	3	0	3	0	
		c_j	7	5	0	0	
$= 21$		Δ_j	0	$-\frac{1}{4}$	0	$\frac{-7}{4}$	

Since all Δ_j's are negative or zero hence the solution is optimal.

$$\therefore \quad Z = 7x_1 + 5x_2,$$

$$Z \max = 7 \times 3 + 5 \times 0$$

$$= 21$$

for $\quad x_1 = 3, x_2 = 0$

$$Z = 21 \quad \text{Ans.}$$

Example 10. Solve

$$\text{Max.} \quad Z = 4x_1 + 10x_2$$

$$\text{s.t.} \quad 2x_1 + x_2 \leq 50$$

$$2x_1 + 5x_2 \leq 100$$

$$2x_1 + 3x_2 \le 90$$

$$x_1, x_2 \ge 0$$

Solution: The equations obtained by introducing slack variables are as follows.

$$2x_1 + x_2 + x_3 = 50$$

$$2x_1 + 5x_2 + x_4 = 100$$

$$2x_1 + 3x_2 + x_5 = 90$$

Taking $x_1 = 0, x_2 = 0$, we get

$$x_3 = 50, x_4 = 100, x_5 = 90$$

which is starting B.F.S.

Starting Simplex Table.

TABLE 1

B	C_B	X_B	Y_1 $(=\alpha_1)$	Y_2 $(=\alpha_2)$	Y_3 $(=\beta_1)$	Y_4 $(=\beta_2)$	Y_4 $(=\beta_3)$	*Mini ratio*
Y_2	0	50	2	1	1	0	0	50
Y_4	0	100	2	$\boxed{5}$	0	1	0	20 mini →
Y_5	0	90	2	3	0	0	1	30
$Z = C_B X_B$		x_j	0	0	50	100	90	
		c_j	4	10	0	0	0	
$= 0$		Δ_j	4	10	0	0	0	
				↑ incoming vector		↓ outgoing vector		

Second Simplex Table

TABLE 2

B	C_B	X_B	Y_1	Y_2	Y_3	Y_4	Y_5	*Mini. ratio*
Y_3	0	30	$\frac{8}{5}$	0	1	$-\frac{1}{5}$	0	
Y_2	10	20	$\frac{2}{5}$	1	0	$\frac{1}{5}$	0	
Y_5	0	30	$\frac{4}{5}$	0	0	$-\frac{3}{5}$	1	
$Z = C_B X_B$		x_j	0	0	30	0	30	
		c_j	4	10	0	0	0	
$= 200$		Δ	0	0	0	-2	0	

Since all Δ_j's are less than or equal to zero, hence the solution is optimal for

$$x_1 = 0, x_2 = 20$$

$$\therefore \quad \text{Max. } Z = 4x_1 + 10x_2$$
$$= 4 \times 0 + 10 \times 20$$
$$= 200 \text{ Ans.}$$

Example 11. Max. $Z = 3x_1 + 2x_2 + 5x_3$

Subject to $x_1 + 2x_2 + x_3 \leq 430$

$3x_1 + 2x_3 \leq 460$

$x_1 + 4x_2 \leq 420$

$x_1, x_2\ x_3 \geq 0$

Solution: The equations obtained by introducing slack variables x_4, x_5, x_6 are as follows.

$$x_1 + 2x_2 + x_3 + x_4 = 430$$
$$3x_1 + 0x_2 + 2x_3 + x_5 = 460$$
$$x_1 + 4x_2 + 0x_3 + x_6 = 420$$

Taking $x_1 = 0, x_2 = 0, x_3 = 0$, we get $x_4 = 430, x_5 = 460, x_6 = 420$ which is starting B.F.S.

Starting Simplex Table

TABLE 1

B	C_B	X_B	Y_1 $(=\alpha_1)$	Y_2 $(=\alpha_2)$	Y_3 $(=\alpha_3)$	Y_4 $(=\beta_1)$	Y_5 $(=\beta_2)$	Y_6 $(=\beta_3)$	*mini. ratio* $\left(\frac{X_B}{Y_1}\right)$
Y_4	0	430	1	2	1	1	0	0	$\frac{430}{1} = 430$
Y_5	0	460	3	0	[2]	0	1	0	$\frac{460}{2} = 230$ →
Y_6	0	420	1	4	0	0	0	1	$\frac{420}{0} = \infty$
$Z = C_B X_B$		x_j	0	0	0	430	460	420	
		c_j	3	2	5	0	0	0	
$= 1150$		Δ_j	3	2	5	0	0	0	

↑ Incoming vector (under Y_3) ↓ Out going vector (under Y_5)

Second Simplex Table

TABLE 2

B	C_B	X_B	Y_1 $(=\alpha_1)$	Y_2 $(=\alpha_2)$	Y_3 $(=\beta_2)$	Y_4	Y_5	Y_6 $(=\beta_2)$	*Mini. ratio* $\left(\frac{X_B}{Y_2}\right)$
Y_4	0	200	$-\frac{1}{2}$	[2]	0	1	$-\frac{1}{2}$	0	$-\frac{5}{3}=100$ →
Y_3	5	230	$\frac{3}{2}$	0	1	0	$\frac{1}{2}$	0	$\frac{230}{0}=\infty$
Y_6	0	420	1	4	0	0	0	1	$\frac{420}{4}$
$Z = C_B X_B$		x_j	0	0	230	200	0	420	
		c_j	3	2	5	0	0	0	
$= 1150$		Δ_j	$-\frac{9}{2}$	2	0	0	$-\frac{5}{2}$	0	
				↑ Incoming vector		↓ Out going vector			

Third Simplex Table

TABLE 3

B	C_B	X_B	Y_1	Y_2 $(=\beta_1)$	Y_3 $(=\beta_2)$	Y_4	Y_5	Y_6 $(=\beta_3)$	*Mini.*
Y_2	2	100	$-\frac{1}{2}$	1	0	$\frac{1}{2}$	$-\frac{1}{4}$	0	
Y_3	5	230	$\frac{3}{2}$	0	1	0	$\frac{1}{2}$	0	
Y_6	0	20	2	0	0	-2	1	1	
$Z = C_B X_B$		x_j	0	100	230	0	0	20	
		c_j	3	2	5	0	0	0	
$= 1350$		Δ_j	-4	0	0	-1	-2	0	

Since all the Δ_j are ≤ 0, the solution $x_1 = 0, x_2 = 100, x_3 = 230$ is optimal with $Z = 1350$ Ans.

Below some more problems have been solved by Simplex method without enumerating the procedures.

Example 12. Solve by Simplex method the L.P.P.

$$\begin{aligned} \text{Max.} \quad & Z = x_1 + x_2 \\ \text{Subject to} \quad & 2x_1 + 3x_2 \leq 22 \\ & 2x_1 + x_2 \leq 14 \\ & x_1 - x_2 \leq 4 \\ & 3x_1 - 2x_2 \geq -6 \\ & x_1, x_2 \geq 0 \end{aligned}$$

Solution: Slack variables are not mentioned separately, and easily understood from the context. Also, non-basic variables are not separately mentioned Using Simplex procedure we get.

TABLE

				1	1	0	0	0	0	*Mini. ratio*
	C_B	B	X_B	Y_1	Y_2	Y_3	Y_4	Y_5	Y_6	
	0	Y_3	22	2	3	1	0	0	0	11
	0	Y_4	14	2	1	0	1	0	0	7
	0	Y_5	4	$\boxed{1}$	–1	0	0	1	0	4 →
	0	Y_6	6	–3	2	0	0	0	1	– ve
			Δ_j	1 ↑	1	0	0	0 ↓	0	
1st Iteration	0	Y_3	14	0	5	1	0	–2	0	14/5
	0	Y_4	6	0	$\boxed{3}$	0	1	–2	0	$\frac{0}{2}$ →
	1	Y_1	4	1	–1	0	0	1	0	–ve
	0	Y_6	18	0	–1	0	0	3	1	–ve
			Δ_j	0	2 ↑	0	0 ↓	–1	0	
2nd Iteration	0	Y_3	4	0	0	1	$-\frac{5}{3}$	$\boxed{\frac{4}{3}}$	0	3 →
								↓		– ve
	1	Y_2	2	0	1	0	$\frac{1}{3}$	$\frac{2}{3}$	0	18
	1	Y_1	6	1	0	0	$\frac{1}{3}$	$\frac{1}{3}$	0	60
	0	Y_6	20	0	0	0	$\frac{1}{3}$	$-\frac{7}{3}$	1	7
			Δ_j	0	0	0 ↓	$-\frac{2}{3}$	$\frac{1}{3}$ ↑	0	
3rd Iteration	0	Y_5	3	0	0	$\frac{3}{4}$	$-\frac{5}{4}$	1	0	
	1	Y_2	4	0	1	$\frac{1}{2}$	$-\frac{1}{2}$	0	0	
	1	Y_1	5	1	0	$-\frac{1}{4}$	$\frac{3}{4}$	0	0	
	0	Y_6	13	0	0	$-\frac{7}{4}$	$\frac{13}{4}$	1	1	
			Δ_j	0	0	$-\frac{1}{4}$	$-\frac{1}{4}$	0	0	

$$Z_{max} = 9 \text{ for } x_1 = 5, x_2 = 4$$
$$Z_{max} = C_B X_B = 0 \times 3 + 1 \times 4 + 1 \times 5 = 9$$
for $x_1 = 5, x_2 = 4$

Example 13. Solve by Simplex method.

Max. $Z = 3x_1 + 2x_2 + x_3$

Subject to $5x_1 + 6x_2 + 10x_3 \le 50$

$5x_1 + 4x_2 \le 20$

$15x_1 + 4x_2 \le 30$

$x_1, x_2, x_3 \ge 0$

Solution: Applying Simplex procedure, we get

TABLE

			3	2	1	0	0	0	*Mini. Ratio*
C_B	B	X_B	Y_1	Y_2	Y_3	Y_4	Y_5	Y_6	
0	Y_4	50	5	6	0	1	0	0	10
0	Y_5	20	5	4	0	0	1	0	4
0	Y_6	30	[15]	4	0	0	0	1	2 →
		Δ_j	3 ↑	2	1	0	0	0 ↓	
0	Y_4	40	0	$\frac{14}{3}$	10	1	0	$-\frac{1}{20}$	$\frac{120}{14}$
0	Y_5	10	0	[$\frac{8}{3}$]	0	0	1	$-\frac{1}{3}$	$\frac{30}{8}$ →
3	Y_6	2	1	$\frac{4}{15}$	0	0	0	$\frac{1}{15}$	$\frac{30}{4}$
		Δ_j	0	↑ $\frac{6}{5}$	1	0	0 ↓	$-\frac{1}{5}$	
0	Y_4	$\frac{45}{2}$	0	0	[10]	1	$-\frac{7}{4}$	$\frac{1}{4}$	$\frac{45}{20}$ →
2	Y_2	$\frac{15}{4}$	0	1	0	0	$\frac{3}{8}$	$-\frac{1}{8}$	∞
3	Y_1	1	1	0	0	0	$-\frac{1}{10}$	$\frac{1}{10}$	∞
		Δ_j	0	0	1 ↑	0 ↓	$-\frac{9}{20}$	$-\frac{1}{20}$	

contd.

1	Y_3	$\frac{9}{4}$	0	0	1	$\frac{1}{10}$	$-\frac{7}{40}$	$\frac{1}{40}$
2	Y_2	$\frac{15}{4}$	0	1	0	0	$\frac{3}{8}$	$-\frac{1}{6}$
3	Y_1	1	1	0	0	0	$-\frac{1}{10}$	$\frac{1}{10}$
		Δ_j	0	0	0	$-\frac{1}{10}$	$-\frac{11}{40}$	$-\frac{3}{40}$

$$Z \text{ max.} = C_B X_B = 1 \times \frac{9}{4} + 2 \times \frac{15}{4} + 3 \times 1$$

$$Z \text{ max.} = \frac{51}{4} \quad \text{at} \quad x_1 = 1$$

$$x_2 = \frac{15}{4} \quad \text{and} \quad x_3 = \frac{9}{4}$$

Example 14. Solve by Simplex Method

Max. $Z = x_1 + x_2$

Subject to $-x_1 + 2x_2 \leq 8$

$3x_1 + x_2 \leq 11$

$x_1, x_2 \geq 0$

Solution: Employing the procedure of Simplex, we obtain

TABLE

			1	1	0	0	
C_B	B	X_B	Y_1	Y_2	Y_3	Y_4	Mini. ratio
0	Y_3	8	–1	[2]	1	0	$\frac{8}{2} \rightarrow 4$
0	Y_4	11	3	1	0	0	$\frac{11}{1}$
			1	1 ↑	0 ↓	0	
1	Y_2	4	$-\frac{1}{2}$	1	$\frac{1}{2}$	0	– ve
0	Y_4	7	[$\frac{7}{2}$]	0	$-\frac{1}{2}$	1	2 →
			$\frac{3}{2}$ ↑	0	$\frac{1}{2}$	0 ↓	
1	Y_2	5	0	1	$\frac{3}{7}$	$\frac{1}{7}$	
1	Y_1	2	1	0	$-\frac{1}{7}$	$\frac{2}{7}$	
		Δ_j	0	0	$-\frac{2}{7}$	$-\frac{3}{7}$	

$Z_{max.} = 7$ at $x_1 = 2, x_2 = 5$

Note—At the first iteration there is a tie and it can be broken arbitrarily.

Example 15. Solve by simplex method.

Max. $Z = 13x_1 + 30x_2$

Subject to $x_1 - x_2 \leq 3$

$3x_1 + x_2 \leq 11$

$x_1 + x_2 \leq 7$

$x_1, x_2 \geq 0$

Solution: Using simplex method, we find

TABLE

C_B	B	X_B	Y_1	Y_2	Y_3	Y_4	Y_4	Mini. ratio
0	Y_3	3	1	−1	1	0	0	−ve
0	Y_4	11	3	1	0	1	0	
0	Y_5	7	1	[2]	0	0	1	$\frac{7}{2}$ →
			13	30 ↑	0	0	0 ↓	
0	Y_3	$\frac{13}{2}$	$\frac{3}{2}$	0	1	0	$\frac{1}{2}$	
0	Y_4	$\frac{15}{2}$	$\frac{5}{2}$	0	0	1	$-\frac{1}{2}$	
30	Y_2	$\frac{7}{2}$	$\frac{1}{2}$	1	0	0	$\frac{1}{2}$	
		Δ_j	−2	0	0	0	−15	

$$Z_{max} = 105 \text{ at } x_1 = 0, x_2 = \frac{7}{2}$$

$$Z_{max} = C_B \times B = 0 \times \frac{13}{2} + 0 \times \frac{15}{2} + 30 \times \frac{7}{2}$$

$$= 105$$

for $x_1 = 0$

$$x_2 = \frac{7}{2}$$

Example 16. Solve by simplex method

$$\begin{aligned} \text{Max.} \quad & Z = 5x_1 + 3x_2 \\ \text{Subject to} \quad & x_1 + x_2 \le 2 \\ & 5x_1 + 2x_2 \le 10 \\ & 3x_1 + 8x_2 \le 12 \\ & x_1, x_2 \ge 0 \end{aligned}$$

[Roorkee, '76]

Solution: Adopting simplex procedure, we have

TABLE

C_B	B	X_B	Y_1	Y_2	Y_3	Y_4	Y_5	*Mini ratio*
0	Y_3	2	1	1	1	0	0	2
0	Y_4	10	$\boxed{5}$	2	0	1	0	2 →
0	Y_5	12	3	8	0	0	1	4
		Δ_j	5 ↓	3	0 ↓	0	0	
0	Y_3	0	0	$\boxed{\frac{3}{5}}$	1	$-\frac{1}{5}$	0	0 →
5	Y_1	2	1	$\frac{2}{3}$	0	$\frac{1}{5}$	0	5
0	Y_5	6	0	$\frac{34}{5}$	0	$-\frac{3}{5}$	1	$\frac{15}{17}$
		Δ_j	0	1 ↑	0 ↓	−1	0	
3	Y_2	0	0	1	$\frac{5}{3}$	$-\frac{1}{3}$	0	
5	Y_1	2	1	0	$-\frac{2}{3}$	$\frac{1}{3}$	0	
0	Y_5	6	0	0	$-\frac{34}{3}$	$\frac{5}{3}$	1	
		Δ_j	0	0	$-\frac{5}{3}$	$-\frac{2}{3}$	0	

$$Z_{\max} = 10 \text{ at } x_1 = 2, x_2 = 0$$

Artificial Variable Technique : Two phase method.

Now shall discuss problems of linear programming where any of the constraints may have (≥) and (=) signs with all br's positive. In such problems we introduce surplus variables in the constraints with sing ≥ and =.

We add one more variables to each of such constraints. These variables are called 'artificial variables'.

The artificial variable technique is merely a device to get the starting B.F.S. So that we can start with simplex procedure until the optimal solution is obtained. Such problems are solved by two methods.

Method I. Two Phase Method.

Phase I. We first eliminate the artificial variables one by one from the simplex table and introduce other variables (non-artificial variables). This part of the solutions in which the artificial variables are removed is called phase I.

Phase II. We proceed to get the optimal solution by usual simplex outline called phase II.

For the reason, this method is called the "Two Phase Method".

Method II. Big M-Method (Carsers M-method). R.U. 83 M.Sc.

The idea of Big M-Method can be explained as follows

(i) We express the problem in the standard form.

(ii) Then add non-negative artificial variables to the left hand side of each of the equations corresponding to constraints of the type $\geq$ and (=). A sign a very large price (or per unit penalty) to the variables in the objective function. Such large price will be denoted by —M for maximization problems (and + M for minimization problem) $M > 0$.

(iii) We use the artificial variable for the starting solution and then proceed with the usual simplex method until the optimal solution is obtained.

The above concept of Artificial Variable Technique are Big- M method can be understood by the following examples.

Example 17. Solve the following problem by two phase method.

$$\text{Mini. } Z = x_1 + x_2$$

$$\text{Subject to} \quad 2x_1 + x_2 \geq 4$$

$$x_1 + 7x_2 \geq 7$$

$$x_1, x_2 \geq 0$$

[Delhi '73, Kurk '75, Raj L.P. '79, '87]

Solution: The problem is a problem of minimization. So, for our convenience, we will convert this problem of minimization to maximization by taking the objective function as

$$z' = -z$$

$$\text{Max.} \qquad z' = -z = -x_1 - x_2$$

Further, all the b_i (4, 7) are positive.

Introducing the surplus variables x_3 and x_4 the constraints become

$$2x_1 + x_2 - x_3 = 4,$$

$$x_1 + 7x_2 - x_4 = 7,$$

Here, we can not get the starting B.F.S. we introduce the artificial variables (positive variables) x_5 and x_6. Thus the above equations can be written as

$$2x_1 + x_2 - x_3 + x_5 = 4$$
$$x_1 + 7x_2 - x_4 + x_6 = 7$$

Clearly $x_5 < x_3, x_6 < x_4$

Phase I. This phase consists of the removal of artificial variables.

Taking $x_1 = 0, x_2 = 0, x_3 = 0, x_4 = 0$ we get
$x_5 = 4, x_6 = 7.$

Let, A_1 and A_2 denote the artificial column vectors corresponding to artificial variables x_5 and x_6 respectively. We construct the first table as follows.

TABLE 1

	X_B	Y_1	Y_2	Y_3	Y_4	A_1 (β_1)	A_2 (β_2)	
A_1	4	2	1	−1	0	1	0	
A_2	7	1	[7]	0	−1	0	1	→
	x_j	0	0	0	0	4	7	
			↑				↓	

Now, we first remove one by one the each of artificial column vector A_1 and A_2 from the basis matrix.

We can remove A_2 and introduce Y_2 in its place in the basis matrix. For this divide the second row by 7 and then subtract it from the first row. Thus we get the next transformed Table.

TABLE 2

	X_B	Y_1	Y_2 (β_2)	Y_3	Y_4	A_1 (β_1)	A_2*	
A_1	3	$\boxed{\frac{13}{7}}$	0	−1	$\frac{1}{7}$	1	$-\frac{1}{7}$	→
Y_2	1	$\frac{1}{7}$	1	0	$-\frac{1}{7}$	0	$\frac{1}{7}$	
	x_j	0	1	0	0	3	0	
		↑				↓		

Whenever the artificial x_6 becomes zero (non-basic), we forget about it and never consider the corresponding vector A_2 again for re-entry into the basis matrix.

Now, we proceed to remove A_1 and introduce Y_1 in its place in the basis matrix. For this we multiply first row by $\frac{7}{13}$ and subtract $\frac{1}{7}$ times of this new from the second row. Thus we get the following table.

TABLE 3

B	X_B	$Y_1\ (\beta_1)$	$Y_2\ (\beta_2)$	Y_3	Y_4	A_1^x
Y_1	$\frac{21}{13}$	1	0	$-\frac{7}{13}$	$\frac{1}{13}$	$\frac{7}{13}$
Y_2	$\frac{10}{13}$	0	1	$\frac{1}{13}$	$-\frac{14}{91}$	$-\frac{1}{13}$
	x_j	$\frac{21}{13}$	$\frac{10}{13}$	0	0	0

Since, artificial variable x_5 become zero so we forget about A_1 and not consider it again.

Thus we get the following solution in phase I.

$$x_1 = \frac{21}{13}, \quad x_2 = \frac{10}{13}, \quad x_3 = 0, \quad x_4 = 0$$

which is the B.F.S. with which we can proceed to get the optimal solution by simplex method.

Phase II. Now, construct the starting simplex table as follows.

TABLE 4

B	C_B	X_B	$Y_1\ (\beta_1)$	$Y_2\ (\beta_2)$	Y_3	Y_4	*Mini. ratio*
Y_1	-1	$\frac{21}{13}$	1	0	$-\frac{7}{13}$	$\frac{1}{13}$	
Y_2	-1	$\frac{10}{13}$	0	1	$\frac{1}{13}$	$-\frac{14}{91}$	
$Z' = C_B X_B$		x_j	$\frac{21}{13}$	$\frac{10}{13}$	0	0	
		c_j	-1	-1	0	0	
$= \frac{31}{13}$		Δ_j	0	0	$-\frac{6}{13}$	$-\frac{7}{91}$	

$$\Delta_3 = C_3 - C_B Y_3 = 0 - (-1, -1)\left(-\frac{7}{13}, \frac{1}{13}\right)$$

$$= \left(-\frac{7}{13}, \frac{1}{13}\right) = -\frac{6}{13}$$

$$\Delta_4 = C_4 - C_B Y_4 = 0 - (-1, -1)\left(\frac{1}{13} \cdot -\frac{14}{91}\right)$$

$$= -\left(-\frac{1}{13} + \frac{14}{91}\right) = -\frac{7}{91}$$

Since Δ_3 and Δ_4 are negative, the optimal solution is

$$x_1 = \frac{21}{13}, \quad x_2 = \frac{10}{13}$$

$$\text{Min. } Z = -\text{max. } Z' = \frac{31}{13}$$

Example solved by Big-M-Method

Example 18. Max. $Z = -2x_1 - x_2$

Subject to
$$3x_1 + x_2 = 3$$
$$4x_1 + 3x_2 \geq 6$$
$$x_1 + 2x_2 \leq 4$$
$$x_1, x_2 \geq 0$$

Solution: Step I. Introducing the slack, surplus and artificial variables, the system of constraint equation becomes

$$3x_1 + x_2 + x_5 = 3$$
$$4x_1 + 3x_2 - x_2 + x_6 = 6$$
$$x_1 + 2x_2 + x_4 = 4$$

Taking $x_1 = 0, x_2 = 0, x_3 = 0$ we get $x_4 = 4, x_5 = 3, x_6 = 6$.

Step II. Assigning the large negative price'—M' to the artificial variables x_5 and x_6, the objective function becomes

$$\text{Max. } Z = -2x_1 + x_2 + 0x_3 + 0x_4 - Mx_5 - Mx_6$$

The Starting Simplex Table is as follows:

TABLE 1

B	C_B	X_B	Y_1	Y_2	Y_3	Y_4	A_1 (β_1)	A_2 (β_2)	*Mini. ratio* $\frac{X_B}{Y_1}$
A_1	$-M$	3	[3]	1	0	0	1	0	$\frac{3}{3}$ →
A_2	$-M$	6	4	3	-1	0	0	1	$\frac{6}{4}$
Y_4	0	4	1	2	0	1	0	0	$\frac{4}{1}$
$Z = C_B X_B$		x_j	0	0	0	4	3	6	
		c_j	-2	-1	0	0	$-M$	$-M$	
		Δ_j	$-2 + 7M$	$-1 + 4M$	$-M$	0	0	0	

↑ incoming vector (under Y_1) ↓ outgoing vector (under A_2)

$$\Delta_1 = c_1 - C_B\, Y_1 = -\,2 - (M, -M, 0)\,(3, 4, 1)$$
$$= -\,2 - (-\,3M - 4M + 0)$$
$$= 2 + 7M$$
$$\Delta_2 = c_2 - C_B\, Y_2 = -\,1 - (-\,M, -\,M, 0)\,(1, 3, 2)$$
$$= -\,1 - (-\,M - 3 \cdot M + 0)$$
$$= -\,1 + 4M$$
$$\Delta_3 = c_3 - C_B\, Y_3 = 0 - (-\,M, -\,M, 0)\,(0, -1, 0)$$
$$= -\,M$$

Here, Max. $\Delta_j = \Delta_1$

$\therefore$ Y_1 is the incoming vector and by minimum ratio rule A_1 (artificial vector) is the outgoing vector.

$\therefore$ Key element $= y_{11} = a_{11} = 3$

In order to bring Y_1 in place of A_1, divide first row by the key element to get 1, then subtract 4 and 1 times from second and 3rd row respectively.

$\therefore$ The Second Simplex Table is as follows:

TABLE 2

B	C_B	X_B $(=\beta_1)$	Y_1	Y_2	Y_3	Y_4 $(=\beta_3)$	A_1	A_2 $(=\beta_2)$	*Mini. ratio* $\frac{X_B}{Y_2}$
Y_1	-2	1	1	$\frac{1}{3}$	0	0	$\frac{1}{3}$	0	$= 3$
A_2	$-M$	2	0	$\boxed{\frac{5}{3}}$	-1	0	$-\frac{4}{3}$	1	$\frac{2}{\frac{5}{3}} = \frac{6}{5}$ →
Y_4	0	3	0	$\frac{5}{3}$	0	1	$-\frac{1}{3}$	0	$\frac{9}{5}$
		x_j	1	0	0	3	0	2	
		c_j	-2	1	0	0	$-M$	$-M$	
		Δ_j	0	$\frac{-1+5M}{3}$	$-M$	0	$\frac{2-7M}{3}$	0	

↑ incoming vector (under Y_1)

↓ outgoing vector (under A_2)

$$\Delta_2 = c_2 - C_B\, Y_2 = -1 - (-2, -M, 0)\left(\frac{1}{3}, \frac{5}{3}, \frac{5}{3}\right)$$
$$= \frac{-1+5M}{3}$$
$$\Delta_3 = c_3 - C_B\, Y_3 = 0 - (-2, -M, 0)\,(0, -1, 0)$$
$$= -M$$
$$\Delta_5 = c_5 - C_B\, Y_5 = -M - (-2, -M, 0)$$
$$\left(\frac{1}{3}, -\frac{4}{3}, -\frac{1}{3}\right)$$

Since Max. $\Delta_j = \Delta_2$

$\therefore$ Y_2 is the incoming vector and by mini.-ratio rule A_2 is outgoing vector

$$\therefore \text{ Key element} = \frac{5}{3}$$

The Third Simplex Table is as follows:

TABLE 3

B	C_B	X_B	Y_1	Y_2	Y_3	Y_4	$Y_5\,(A_1)$	$Y_6\,(A_3)$	*Mini. ratio*
Y_1	-2	$\frac{3}{5}$	1	0	$\frac{1}{5}$	0	$\frac{3}{5}$	$-\frac{1}{5}$	
Y_2	-1	$\frac{6}{5}$	0	1	$-\frac{3}{5}$	0	$-\frac{4}{5}$	$\frac{3}{5}$	
Y_4	0	1	0	0	1	1	1	-1	
$Z = C_B X_B$		x_j	$\frac{3}{5}$	$\frac{6}{5}$	0	1	0	0	
$= \frac{-12}{5}$		c_j	-2	-1	0	0	$-M$	$-M$	
		Δ_j	0	0	$-\frac{1}{5}$	0	$-M+\frac{2}{5}$	$-M+\frac{1}{5}$	

To test the optimality, we compute

$$\Delta_3 = c_3 - C_B\, Y_3 = 0 - (-2, -1, 0)\left(\frac{1}{5}, -\frac{3}{5}, 0\right)$$
$$= -\frac{1}{5}$$
$$\Delta_5 = c_5 - C_B\, Y_5 = -M - (-2, -1, 0)\left(\frac{3}{5}, -\frac{4}{5}, 1\right)$$
$$= -M + \frac{2}{5}$$
$$\Delta_6 = c_6 - C_B\, Y_6 = -M - (-2, -1, 0)\left(-\frac{1}{5}, \frac{3}{5}, -1\right)$$

$$= -M - \left(\frac{2}{5} - \frac{3}{5}\right)$$

$$= -M - \frac{2}{5} + \frac{3}{5}$$

$$= -M + \frac{1}{5}$$

Since M is as large as possible; Δ_3, Δ_5, Δ_6 are all negative. So, this solution is optimal.

$\therefore$ Optimal solution is for

$$x_1 = \frac{3}{5}, \quad x_2 = \frac{6}{5}$$

$$\text{Max. } z = -2x_1 - x_2$$

$$= -2 \times \frac{3}{5} - \frac{6}{5}$$

$$\text{Max. } z = -\frac{6}{5} - \frac{6}{5} = -\frac{12}{5}$$ Ans.

Apply Charner Big-M method or Two phase method to solve the following problem—

$$\text{Min. } Z = 3x_1 + x_3$$

$$\text{Subject to } 2x_1 + x_2 \geq 14$$

$$x_1 - x_2 \geq 4$$

$$x_1, x_2 \geq 0$$

Solution: Substituting surplus variable x_3, x_4 and artificial variable x_5, x_6 to the constraints, we get

$$2x_1 + x_2 - x_3 + x_5 = 14$$

$$x_1 - x_2 - x_4 + x_6 = 4$$

$$x_1, \; x_2, x_3, x_4 \geq 0$$

and $$x_5, x_6 \geq 0$$

Also, the objective function becomes with M a large positive number.

$$\text{Max. } Z = -3x_1 - x_2 + 0.x_3 + 0.x_4$$

$$= -Mx_5 - Mx_6$$

By employing the procedure, we get the following table.

TABLE

C_B	B	X_B	Y_1	Y_2	Y_3	Y_4	A_1	A_2
$-M$	A_1	14	2	1	-1	0	1	0
$-M$	A_2	4	$\boxed{1}$	-1	0	-1	0	1
		Δ_j	$\uparrow$ $3M-3$	-1	$-M$	$-M$	0	0
$-M$	A_1	6	0	$\boxed{3}$	-1	2	1	-2
-3	Y_1	4	1	-1	0	-1	0	1
		Δ_j	0	$3M-4$ $\uparrow$	$-M$	$2M-3$	0	$-3M-3$
-1	Y_2	2	0	1	$-\frac{1}{3}$	$\frac{2}{3}$	$\frac{1}{3}$	$-\frac{2}{3}$
-3	Y_1	6	1	0	$-\frac{1}{3}$	$-\frac{1}{3}$	$\frac{1}{3}$	$\frac{1}{3}$
$Z = C_B X_B =$ $-1\times2 + -3\times6 = -20$		Δ_j	0	0	$-\frac{4}{3}$	$\frac{1}{3}$	$\frac{4}{3} - M$	$\frac{1}{3} - M$

Z min. = 20 at $x_1 = 6$, $x_2 = 2$

Example 20. Solve the following L.P.P.

Max. $z = 3x_1 + 2x_2 + x_3$

Subject to

$$-3x_1 + 2x_2 + 2x_3 = 8$$
$$-3x_1 + 4x_2 + x_3 = 7$$
$$x_1, x_2, x_3 \geq 0$$

Solution: Clearly there is no need of introducing slack or surplus variables to the constraints. So we introduce the artificial variables x_4 and x_5. Then given L.P.P. reduces to the following form.

Max. $z = 3x_1 + 2x_2 + x_3 - Mx_4 - Mx_5$

$$-3x_1 + 2x_2 + 2x_3 + x_4 + 0x_5 = 8$$
$$-3x_1 + 4x_2 + x_3 + 0x_4 + x_5 = 7$$
$$x_1, x_2, x_3, x_4, x_5 \geq 0$$

Putting $x_1 = x_2 = x_3 = 0$, we get $x_4 = 8$, $x_5 = 7$ as starting B.F.S.

All computation work is done in the following table without enumerating the procedures.

TABLE

		c_j	3	2	1	$-M$	$-M$	Mini. ratio
B	C_B	X_B	Y_1	Y_2	Y_3	A_1	A_2	$\frac{X_B}{Y_2}$
A_1	$-M$	8	-3	2	2	1	0	4
A_2	$-M$	7	-3	$\boxed{4}$	1	0	1	$\frac{7}{4} \rightarrow$
$Z = C_B X_B = -15M$		Δ_j	$-6M + 3$	$6M + 2$	$3M + 1$	0	0	
				$\uparrow$			$\downarrow$	$\frac{X_B}{Y_3}$
A_1	$-M$	$\frac{9}{2}$	$-\frac{3}{2}$	0	$\boxed{\frac{3}{2}}$	1		$3 \rightarrow$
Y_2	2	$\frac{7}{4}$	$-\frac{3}{4}$	1	$\frac{1}{4}$	0		7
$Z = C_B X_B = 7 - \frac{9M}{2}$		Δ_j	$\frac{-3M+9}{2}$	0	$\frac{3M+1}{2}$	0		
					$\uparrow$	$\downarrow$		$\frac{X_B}{Y_1}$
Y_3	1	3	-1	0	1			$-$ ve
Y_2	2	1	$-\frac{1}{2}$	1	0			$-$ ve
		Δ_j	5	0	0			
			$\uparrow$					

We see that Y_1 is incoming vector, But we can not select the out giving vector as in the minimum ratio colours all elements are negative. Hence in the case the solution is unbounded.

Example 21. Solve the following L.P.P.

Max. $z = x_1 + 2x_2 + 3x_3 - x_4$

Subject to

$$x_1 + 2x_2 + 3x_3 = 15$$
$$2x_1 + x_2 + 5x_3 = 20$$
$$x_1 + 2x_2 + x_3 + x_4 = 10$$
$$x_1, x_2, x_3, x_4 \geq 0$$

[Meerut 71, 77, 81, 82 (P), L.P. 87, 88, Raj L.P. 80, 86, 88, Gauhati M. Sc. (Stat 75)]

Solution: Obviously we are not required to introduce slack or surplus variable we introduce the artificial variables x_5 and x_6 and assign large negative price $-M$ to the artificial variables. Then the LPP takes the following form.

Max. $z = x_1 + 2x_2 + 3x_3 - x_4 - Mx_5 - Mx_6$

Subject to

$$x_1 + 2x_2 + 3x_3 + 0x_4 + x_5 = 15$$
$$2x_1 + x_2 + 5x_3 + 0x_4 + 0x_5 + x_6 = 20$$
$$x_1 + x_2 + x_3 + x_4 = 10$$
$$x_1, x_2, x_3, x_4, x_5, x_6 \geq 0$$

Putting $x_1 = x_2 = x_3 = 0$ we get $x_4 = 10, x_5 = 15, x_6 = 20$ as starting B.F.S.

All computation is shown in the following tables.

TABLE

		c_j	1	2	3	–1	$-M$	$-M$	*Mini. Ratio*
B	C_B	X_B	Y_1	Y_2	Y_3	Y_4	A_1	A_2	$\frac{X_B}{Y_2}$
A_1	$-M$	15	1	2	3	0	1	0	5
A_2	$-M$	20	2	1	[5]	0	0	1	4→
A_4	1	10	1	2	1	1	0	0	10
$Z = C_B X_B$ $= -35M - 10$		Δ_j	$3M + 2$	$3M + 4$	$8M+4$	0	0	0	
					↑				$\frac{X_B}{Y_3}$
A_1	$-M$	3	$-\frac{1}{5}$	[$\frac{7}{5}$]	0	0	1		$\frac{15}{7}$→
Y_3	3	4	$\frac{2}{5}$	$\frac{1}{5}$	1	0	0		4
Y_4	–1	6	$\frac{3}{5}$	$\frac{9}{5}$	0	1	0		$\frac{10}{3}$
$Z = C_B X_B$ $6 - 3M$		Δ_j	$\frac{2-M}{5}$	$\frac{75+16}{5}$	0	0	0		
				↑					$\frac{X_B}{Y_1}$
Y_2	2	$\frac{15}{7}$	$-\frac{1}{7}$	1	0	0	– 15		
Y_3	3	$\frac{25}{7}$	$\frac{3}{7}$	0	1	0	$\frac{25}{3}$		

contd.

Y_4 −1	$\frac{15}{7}$	$\boxed{\frac{6}{7}}$	0	0	1	$\frac{15}{6}$ →	
$Z = C_B X_B = \frac{90}{7}$	Δ_j	$\frac{6}{7}$ ↑	0	0	0		

		c_j	1	2	3	−1	−M	−M	
B	C_B	X_B	Y_1	Y_2	Y_3	Y_4	A_1	A_2	Mini. Ratio
Y_2	2	$\frac{5}{2}$	0	1	0	$\frac{1}{6}$			
Y_3	3	$\frac{5}{2}$	0	0	1	$-\frac{1}{2}$			
Y_1	1	$\frac{5}{2}$	1	0	0	$\frac{7}{6}$			
$Z = C_B X_B$		Δ_j	0	0	0	1			

Max $\quad Z = 2 \times \frac{5}{2} \times 3 \times \frac{5}{2} + 1 \times \frac{5}{2} = 15$

Here $\quad \Delta_j \leq 0, \forall j$

$\therefore \quad Z = 15$ is optimal solution Z max 15

for $\quad x_1 = \frac{5}{2}, x_2 = \frac{5}{2}, x_3 = \frac{5}{2}$

Example 22. Use two phase simplex method to solve

Max. $\quad Z = 3x_1 + 2x_2$

Subect to $\quad 2x_1 + x_2 \leq 2$

$3x_1 - 4x_2 \geq 12$

$x_1, x_2 \geq 0$

Introducing a slack variable $x_3 \geq 0$ and a surplus variable $x_4 \geq 0$, the given constraints can be converted into the following equations

$$2x_1 + x_2 + x_3 = 2$$
$$3x_1 + 4x_2 - x_4 = 12$$

However to obtain a unit matrix and consequently an initial basic feasible solution to the problem, we need an artificial variable x_5 to be introduced in the second equation.

Thus we have,

$$2x_1 + x_2 + x_3 = 2$$
$$3x_1 + 4x_2 - x_4 + x_5 = 12$$

Putting $x_1 = 0$, $x_2 = 0$, $x_4 = 0$, we get an initial basic feasible solution $x_3 = 2$, $x_5 = 12$.

Phase I: Now, following rule we write the simplex table

Phase II: (It has to be remembered that we assign a cost—I each to the artificial variables and cost 0 to all other variables in Phase I.)

B	C_B	X_B	Y_1	Y_2	Y_3 (β_1)	Y_4	A_1 (β_2)	Mini. ratio $\frac{X_B}{Y_2}$
Y_3	0	2	2	[1]	1	0	0	$\frac{2}{1} = 2$
A_1	– 1	12	3	4	0	–1	1	$\frac{12}{4} = 3$
$Z = C_B X_B$		x_j	0	0	2	0	12	
= – 12		c_j	0	0	0	0	–1	
		Δ_j	3	4	0	–1	0	
				↑	↓			

Here
$$\Delta_1 = c_1 - C_B\, Y_1 = 0 - (-3) = 3$$
$$\Delta_2 = c_2 - C_B\, Y_2 = 0 - (-4) = 4$$
$$\Delta_4 = c_4 - C_B\, Y_4 = -1$$

Now max. $\Delta_j = \Delta_2$.

Hence Y_2 is the incoming vector.
By min. ratio rule, we find that β_1 is the out going vector.

$\therefore$ I is the key element.

Now we make the following intermediate table, as before

X_B	Y_1	Y_2	Y_3	Y_4	A_1
2	2	1	1	0	0
4	–5	0	–4	–1	1

Hence the second simplex table is

B	C_B	X_B	Y_1	Y_2 (β_1)	Y_3	Y_4	A_1 (β_2)
Y_2	0	2	2	1	1	0	0
A_1	–1	4	–5	0	–4	–1	1
$Z = C_B\, X_B = -4$		x_j	0	2	0	0	4
		c_j	0	0	0	0	–1
		Δ_j	–5	0	–4	–1	0

Since all $\Delta_j \leq 0$, optimal basic feasible solution to the auxiliary LP problem has been attained. But at the same time $Z < 0$ and the artificial vector A_1 appears in the basic at a positive level (since $x_5 = 4$).

Hence the given LP problem does not possess any feasible solution.

Example 23. Solve by simplex

Max. $Z = 9x_1 + 7x_2$

Subject to

$$x_1 + 2x_2 \leq 7$$
$$x_1 - x_2 \leq 4$$
$$x_1, x_2 \geq 0.$$

Solution: Introducing the slack variables x_3 and x_4 to convert the inequalities into equations, we have

$$x_1 + 2x_2 + x_3 = 7 \quad ...(i)$$
$$x_1 - x_2 + x_4 = 4 \quad ...(ii)$$

The equations (i) and (ii) can be rearranged as

$$x_1 + 2x_2 + x_3 + 0x_4 = 7 \quad ...(iii)$$
$$x_1 - x_2 + 0x_3 + x_4 = 4 \quad ...(iv)$$

and objective function as

$$Z = 9x_1 + 7x_2 + 0.x_3 + 0.x_4 \quad ...(v)$$

Taking $x_1 = 0$ and $x_2 = 0$ in (iii) and (iv) we get

$$x_3 = 7 \text{ and } x_4 = 4$$

Thus the starting B.F.S. is

$$x_1 = 0, x_2 = 0, x_3 = 7 \text{ and } x_4 = 4.$$

SIMPLEX TABLE 1

B	C_B	X_B	Y_1	Y_2	Y_3	Y_4	Min. Ratio $= X_B$/Incoming vector
Y_3	0	7	1	2	1	0	$\frac{7}{1} = 7$
Y_4	0	4	[1]	–1	0	1	$\frac{4}{1} = 4$ →
$Z = C_B X_B$ $= (0, 0) \times (7, 4)$ $= 0 \times 7 + 0 \times 4$ $= 0$		x_j c_j Δ_j	0 9 9	0 7 7	7 0 0	4 0 0	
			↑			↓	

Where $\Delta_j = C_j - C_B Y_j$

$\therefore$ $\Delta_1 = c_1 - C_B Y_1 = 9 - (0, 0)(1, 1) = 9 - 0 \times 1 + 0 \times 1 = 9 - 0 = 9$

$\Delta_2 = c_2 - C_B Y_2 = 7 - (0, 0)(2, -1) = 7 - (0 \times 2 + 0 \times - 1) = 7 - 0 = 7$ and so on.

Since $\Delta_j \leq 0.\ \forall_j$, therefore $Z = 0$ is not optimal, we proceed to form simplex table II as

SIMPLEX TABLE II

B	C_B	X_B	Y_1	Y_2	Y_3	Y_4	Mini. Ratio $= X_B/$Incoming vector
Y_3	0	3	0	[3]	1	−1	$\frac{3}{3} = 1$ →
Y_1	9	4	1	−1	0	1	$\frac{4}{-1} = -4$ (neglected)
$Z = C_B X_B$ $= 0 \times 2 + 9 \times 4$ $= 36$		x_j c_j Δ_j	4 9 0	0 7 16	3 0 0	0 0 −9	
				↑	↓		

Since all Δ_j are neither zero nor negative, we proceed to from table (III) as

SIMPLEX TABLE III

B	C_B	X_B	Y_1	Y_2	Y_3	Y_4	Mini. Ratio $= X_B/$Incoming vector
Y_2	7	1	0	1	$\frac{1}{3}$	$-\frac{1}{3}$	
Y_1	9	5	1	0	$\frac{1}{3}$	$\frac{2}{3}$	
$Z = C_B X_B$ $= 7 \times 1 + 9 \times 5$ $= 52$		x_j c_j Δj	5 9 0	1 7 0	0 0 $-\frac{16}{3}$	0 0 $-\frac{11}{3}$	

So $\quad Z$ max. $= 52$ for $x_1 = 5$ and $x_2 = 1$

$$\text{Max. } Z = x_1 + 7x_2$$

Subject to

$$-x_1 + 2x_2 \le 8$$

$$x_1 - x_2 \le 4$$

$$x_1, x_2 \ge 0$$

Introducing the slack variables x_3 and x_4 to convert the inequalities into equations we have

$$-x_1 + 2x_2 + x_3 = 8$$

$$\text{and} \quad x_1 - x_2 + x_4 = 4$$

Above equations can be rearranged as

$$-x_1 + 2x_2 + x_3 + 0.x_4 = 8 \quad \text{...(i)}$$

$$x_1 - x_2 + 0.x_3 + x_4 = 4 \quad \text{...(ii)}$$

and the objective function in general form is

$$Z = x_1 + 7x_2 + 0.x_3 + 0.x_4 \quad \text{...(iii)}$$

Putting $x_1 = 0$ and $x_2 = 0$ in (i) we get

$$x_3 = 8 \text{ and } x_4 = 4$$

Thus the starting basic feasible solution (B.F.S.) is

$$x_1 = 0, x_2 = 0, x_3 = 8 \text{ and } x_4 = 4$$

SIMPLEX TABLE I

B	C_B	X_B	Y_1	Y_2	Y_3	Y_4	Minium Ratio = X_B/ Incoming vector
Y_3	0	8	–1	$\boxed{2}$	1	0	8/2 = 4 →
Y_4	0	4	1	–1	0	1	–4 (neglected)
$Z = C_B X_B$ = (0, 0) × (8, 4) = 0 × 8 + 0 × 4 = 0		x_j c_j Δ_j	$x_1 = 0$ $c_1 = 1$ $\Delta_1 = 1$	$x_2 = 0$ $c_2 = 7$ $\Delta_2 = 7$	$x_3 = 8$ $c_3 = 0$ $\Delta_3 = 0$	$x_4 = 4$ $c_4 = 0$ $\Delta_4 = 0$	
				↑	↓		

where, $\Delta_j = c_j - C_B Y_j$

$$\therefore \quad \Delta_1 = c_1 - C_B Y_1 = 1 - (0, 0)(-1, 1) = 1 - (0 \times -1 + 0 \times 1) = 1 - 0$$

$$\therefore \quad \Delta_1 = 1$$

$$\Delta_2 = c_2 - C_B Y_2 = 7 - (0, 0)(2, -1) = 7 - 0$$

$$\therefore \quad \Delta_2 = 7$$

and so on for Δ_3 and Δ_4.

Here in simplex table (I) we do not get all Δ_j either '– ve' or 'zero' so we proceed further to from next simplex table no. (II).

SIMPLEX TABLE II

B	C_B	X_B	Y_1	Y_2	Y_3	Y_4	Minium Ratio $= X_B/$ Incoming vector
Y_2	7	4	$\frac{-1}{2}$	1	$\frac{1}{2}$	0	$\frac{4}{\frac{-1}{2}} = -8$
Y_4	0	8	$\boxed{\frac{1}{2}}$	0	$\frac{1}{2}$	1	$\frac{8}{\frac{1}{2}} = 16$ →
$Z = C_B X_B$		x_j	0	4	0	8	
$= 7 \times 4 + 0 \times 8$		c_j	1	7	0	0	
$= 28$		Δ_j	$\frac{9}{2}$	0	$\frac{-7}{2}$	0	
			↑			↓	

$$\Delta_1 = c_1 - C_B Y_1 = 1 - (7, 0), \left(\frac{-1}{2}, \frac{1}{2}\right)$$

$$= 1 - \left(7 \times \frac{-1}{2} + 0 \times \frac{1}{2}\right)$$

$$= 1 - \left(-\frac{7}{2} + 0\right)$$

$$= 1 + \frac{7}{2} = \frac{9}{2} \text{ and so on.}$$

We need to form another simplex table as all Δ_j are not –ve or zero.

SIMPLEX TABLE III

B	C_B	X_B	Y_1	Y_2	Y_3	Y_4	Minium Ratio $= X_B/$ Incoming vector
Y_2	7	12	0	1	1	1	
Y_1	1	16	1	0	1	2	
$Z = C_B X_B$		x_j	16	12	0	0	
$= 7 \times 12 + 1 \times 16 = 100$		c_j Δ_j	1 0	7 0	0 –8	0 –9	

Z max. $= 100$ for $x_1 = 16$, $x_2 = 12$ Ans.

Example 25. Max. $Z = 10x_1 + x_2 + 2x_3$

Subject to
$$x_1 + x_2 - 2x_3 \le 10$$
$$4x_1 + x_2 + x_3 \le 20$$
$$x_1, x_2, x_3 \ge 0$$

Introducing slack variables x_4 and x_5 to change the inequalities into equations, we have

$$x_1 + x_2 - 2x_3 + x_4 = 10$$
$$4x_1 + x_2 + x_3 + x_5 = 20$$

These equations can be put as

$$x_1 + x_2 - 2x_3 + x_4 + 0.x_5 = 10 \quad ...(i)$$
$$4x_1 + x_2 + x_3 + 0.x_4 + x_5 = 20 \quad ...(ii)$$

and the objective function in general form is

$$Z = 10x_1 + x_2 + 2x_3 + 0.x_4 + 0.x_5 \quad ...(iii)$$

Putting $x_1 = 0$, $x_2 = 0$ and $x_3 = 0$ in (i) and (ii) we get

$$x_4 = 10 \text{ and } x_5 = 20$$

So the basic Feasible is

$$x_1 = 0, x_2 = 0, x_3 = 0, x_4 = 10 \text{ and } x_5 = 20$$

SIMPLEX TABLE I

B	C_B	X_B	Y_1	Y_2	Y_3	Y_4	Y_5	Mini. Ratio = XB/Incoming Vector
Y_4	0	10	1	1	–2	1	0	10/1 = 10
Y_5	0	20	[4]	1	1	0	1	20/4 = 5 →
$Z = C_B X_B$ = 0 × 10 + 0 × 20 = 0		x_j c_j Δ_j	0 10 10	0 1 1	0 2 2	10 0 0	20 0 0	
			↑				↓	

$$\Delta_1 = c_1 - C_B y_1$$
$$= 10 - (0, 0)\,(1, 4) = 0 - (0 \times 1 + 0 \times 4) = 10 - 0 = 10$$

and so on.

SIMPLEX TABLE II

B	C_B	X_B	Y_1	Y_2	Y_3	Y_4	Y_5	*Mini. Ratio*
Y_4	0	5	0	$\frac{3}{4}$	$-\frac{9}{4}$	0	$-\frac{1}{4}$	
Y_1	10	5	1	$\frac{1}{4}$	$\frac{1}{4}$	0	$\frac{1}{4}$	
$Z = C_B X_B$ $= 0 \times 5 +$ 10×5 $= 50$		x_j c_j Δ_j	5 10 0	0 1 $-3/2$	0 2 $-1/2$	5 0 0	0 0 $-5/2$	

So, Z max = 50 for $x_1 = 5, x_2 = 0$ and $x_3 = 0$ Ans.

Example 26. **Max $Z = 3x_1 + 2x_2$**

Subject to,
$$2x_1 + x_2 \le 10$$
$$x_1 + 3x_2 \le 6$$
$$x_1, x_2 \ge 0$$

Solution: Introducing slack variables x_3 and x_4 we have
$$2x_1 + x_2 + x_3 = 10$$
$$x_1 + 3x_2 + x_4 = 6$$

In generalised form the constraints are
$$2x_1 + x_2 + x_3 + 0.x_4 = 10 \quad \text{...(i)}$$
$$x_1 + 3x_2 + 0.x_3 + x_4 = 6 \quad \text{...(ii)}$$
and the objective function in generalised form is
$$Z = 3x_1 + 2x_2 + 0.x_3 + 0.x_4 \quad \text{...(iii)}$$

Putting $x_1 = 0, x_2 = 0$ in (i) and (ii) we have
$$x_3 = 10 \text{ and } x_4 = 6$$
So, the starting B.F.S. are
$$x_1 = 0, x_2 = 0, x_3 = 10 \text{ and } x_4 = 6$$

SIMPLEX TABLE 1

B	C_B	X_B	Y_1	Y_2	Y_3	Y_4	*Mini. Ratio = X_B/ Incoming vector*
Y_3	0	10	[2]	1	1	0	10/2 = 5 →
Y_4	0	6	1	3	0	1	6/1 = 6
$Z = C_B \times B$ $= 0 \times 10 +$ 0×6 $= 0$		x_j c_j Δ_j	0 3 3	0 2 2	10 0 0	6 0 0	
			↑		↓		

$\Delta_1 = C_1 - C_B Y_1 = 3 - (0, 0)(2, 1) = 3 - (0 \times 2 + 0 \times 1) = 3 - 0 = 3$ and so on for Δ_2, Δ_3 and Δ_4.

SIMPLEX TABLE 2

B	C_B	X_B	Y_1	Y_2	Y_3	Y_4	*Mini. Ratio* $= X_B$/ *Incoming vector*
Y_1	3	5	1	$\frac{1}{2}$	$\frac{1}{2}$	0	= 10
Y_4	0	1	0	$\boxed{\frac{5}{2}}$	$-\frac{1}{2}$	1	$\frac{2}{5}$ →
$Z = C_B X_B$		x_j c_j Δ_j	5 3 0	0 2 1/2	0 0 –3/2	1 0 0	
				↑		↓	

$\Delta_1 = c_1 - C_B Y_1 = 3 - (3, 0), (1, 0) = 3 - (3 \times 1 + 0 \times 0) = 3 - 3 = 0.$

SIMPLEX TABLE 3

B	C_B	X_B	Y_1	Y_2	Y_3	Y_4	*Mini. Ratio* $= X_B$/ *Incoming vector*
Y_1	3	$\frac{24}{5}$	1	0	$\frac{3}{5}$	$\frac{-2}{5}$	
Y_2	2	$\frac{2}{5}$	0	1	$\frac{-1}{5}$	$\frac{7}{5}$	
$Z = C_B X_B$ $= 3 \times \frac{24}{5}$ $+ 2 \times \frac{2}{5}$ $= \frac{76}{5}$		x_j c_j Δ_j	$\frac{24}{5}$ 3 0	$\frac{2}{5}$ 2 0	0 0 $\frac{-7}{5}$	0 0 $-\frac{2}{5}$	

So, $\quad$ Max. $Z = \frac{76}{5}$ for $x_1 = \frac{24}{5}$ and $x_2 = \frac{2}{5}$ **Ans.**

Example 27. Max. $R = 3x + 7y + 6z$

Subject to $\quad 2x + 2y + 2z \le 8$

$$x + y \le 3$$
$$y + z \le 6$$
$$x, y, z \ge 0 \qquad \text{[R.U.B.Sc. 85]}$$

Solution: Introducing the slack variables x_4, x_5 and x_6 we get

$$2x_2 + 2y + 2z + x_4 = 8$$
$$x + y + x_5 = 3$$
$$y + z + x_6 = 6$$

In rearranged form the constraints are

$$2x + 2y + 2z + x_4 + 0x_5 + 0x_6 = 8 \qquad ...(i)$$
$$x + y + 0.z + 0.x_4 + x_5 + 0x_6 = 3 \qquad ...(ii)$$
$$0.x + y + z + 0.x_4 + 0.x_5 + x_6 = 6 \qquad ...(iii)$$

and objective function is

$$R = 3x + 7y + 6z + 0.x_4 + 0.x_5 + 0.x_6 \qquad ...(iv)$$

Putting the decision variables $x = 0, y = 0, z = 0$ we get from eqns. (i), (ii) and (iii)
The starting B.F.S.

$$x = 0, y = 0, z = 0, x_4 = 8, x_5 = 3, x_6 = 6$$

SIMPLEX TABLE 1

B	C_B	X	Y_1 *coff of* x	Y_2 *coff of y*	Y_3 *coff of z*	Y_4 *coff of* x_4	Y_5 *coff of* x_5	Y_6 *coff of* x_6	*Min. Ratio* = X_B/ *Incoming vector*
Y_4	0	8	2	2	2	1	0	0	8/2 = 4
Y_5	0	3	1	[1]	0	0	1	0	3/1 = 3 →
Y_6	0	6	0	1	1	0	0	1	6/1 = 6
$R = C_B X_B$ = 0 × 8 + 0 × 3 + 0 × 6 = 0	x_j c_j Δ_j		0 3 3	0 7 7	0 6 6	8 0 0	3 0 0	6 0 0	
				↑			↓		

$$\Delta_1 = c_1 - C_B Y_1 = 3 - (0, 0, 0) \cdot (2, 1, 0)$$
$$= 3 - (0 \times 2 + 0 \times 1 + 0 \times 0) = 3 - 0 \quad \therefore \quad \Delta_1 = 3$$

and so on for Δ_2, Δ_3, Δ_4, Δ_5 and Δ_6

SIMPLEX TABLE 2

B	C_B	X_B	Y_1	Y_2	Y_3	Y_4	Y_5	Y_6	*Min. Ratio*
Y_4	0	2	0	0	[2]	1	–2	0	2/2 = 1 →
Y_2	7	3	1	1	0	0	1	0	3/0 (neglect)
Y_6	0	3	–1	0	1	0	– 1	1	3/1 = 3
$R = C_B X_B$ = 0 × 2 + 7 × 3 + 0 × 3 = 21	x_j c_j Δ_j	0 3 –4	3 7 0	0 6 6	2 0 0	0 0 –7	3 0 0	

↑ (under Y_3) ↓ (under Y_4)

$$\begin{aligned}\Delta_1 &= c_1 - C_B Y_1\\ &= 3 - (0, 7, 0)\,(0, 1, -1)\\ &= 3 - (0 \times 0 + 7 \times 1 + 0 - 1)\\ &= 3 - 7\\ &= -4\end{aligned}$$

$$\begin{aligned}\Delta_3 &= c_3 - C_B Y_3\\ &= 6 - (0, 7, 0)\,(2, 0, 1)\\ &= 6 - (0 \times 2 + 7 \times 0 + 0 \times 1)\\ &= 6 - 0\\ &= 6\end{aligned}$$

and so on.

SIMPLEX TABLE 3

B	C_B	X_B	Y_1	Y_2	Y_3	Y_4	Y_5	Y_6	*Min. Ratio* = X_B/ *Incoming vector*
Y_3	6	1	0	0	1	$\frac{1}{2}$	–1	0	
Y_2	7	3	1	1	0	0	1	0	
Y_6	0	2	–1	0	0	$-\frac{1}{2}$	0	1	
$R = C_B X_B$ = 6 × 1 + 7 × 3 + 0 = 21 + 6 = 27	x_j c_j Δ_j	0 3 –4	3 7 0	1 6 0	0 0 –3	0 0 –1	2 0 0	

So R max. = 27 for $x = 0$, y = 3, $z = 1$ Ans.

Example 28. Max. $Z = x_1 - x_2 + \frac{3}{2} x_3$

Subject to
$$x_1 + x_2 + \frac{3}{2} x_3 \le 10$$
$$2x_2 - x_3 \le 2$$
$$2x_1 - 2x_2 + 3x_3 \le 12$$
$$x_1, x_2, x_3 \ge 0$$

Solution: Introducing slack variables x_4, x_5 and x_6 we get

$$x_1 + x_2 + x_3 + x_4 + 0.x_5 + 0.x_6 = 10 \quad \text{...(i)}$$
$$0.x_1 + 2x_2 - x_3 + 0.x_4 + x_5 + 0.x_6 = 2 \quad \text{...(ii)}$$
$$2x_1 - 2x_2 + 3x_3 + 0.x_4 + 0.x_5 + x_6 = 12 \quad \text{...(iii)}$$

and in general form the objective function can be written as

$$Z = x_1 - x_2 + \frac{3}{2} x_3 + 0x_4 + 0x_5 + 0x_6 \quad \text{...(iv)}$$

Putting the decision variables equal to zero as $x_1 = 0$, $x_2 = 0$ and $x_3 = 0$ in (i), (ii), (iii) we get the value of introduced slack variables as

$x_4 = 10$, $x_5 = 2$ $x_6 = 12$

So starting B.F.S.

$$x_1 = 0, x_2 = 0, x_3 = 0, x_4 = 10, x_5 = 2, x_6 = 12.$$

SIMPLEX TABLE 1

B	C_B	X_B	Y_1	Y_2	Y_3	Y_4	Y_5	Y_6	*Min. Ratio* $= X_B/$ *Incoming vector*
Y_4	0	10	1	1	1	1	0	0	$\frac{10}{1} = 10$
Y_5	0	2	0	2	–1	0	1	0	$\frac{2}{-1} = -2$ (neglected)
Y_6	0	12	2	–2	$\boxed{3}$	0	0	1	$\frac{12}{3} = 4$ →
$Z = C_B X_B$	x_j	0	0	0	10	2	12	
$= 0 \times 10 +$	c_j	1	–1	$\frac{3}{2}$	0	0	0	
$0 \times 2 + 0 \times 12 = 0$	Δ_j	1	–1	$\frac{3}{2}$	0	0	0	
				↑			↓	

$$\Delta_1 = c_1 - C_B Y_1 = 1 - (0, 0, 0) \times (1, 0, 2)$$
$$= 1 - (0 \times 0 + 0 \times 0 + 0 \times 2)$$
$$= 1 - 0 = 1$$

and so on for Δ_2, Δ_3, Δ_4, Δ_5 and Δ_6

SIMPLEX TABLE 2

B	C_B	X_B	Y_1	Y_2	Y_3	Y_4	Y_5	Y_6	*Min. Ratio* $= X_B/$ *Incoming vector*
Y_4	0	6	$\frac{1}{3}$	$\frac{5}{3}$	0	1	0	$-\frac{1}{3}$	
Y_5	0	14	$\frac{2}{3}$	$\frac{4}{3}$	0	0	1	$\frac{1}{3}$	
Y_3	$\frac{2}{3}$	4	$\frac{2}{3}$	$-\frac{2}{3}$	1	0	0	$\frac{1}{8}$	
$Z = C_B X_B$		x_j	0	0	4	6	14	0	
$= \frac{3}{2} \times 4 + 0$		c_j	1	-1	$\frac{3}{2}$	0	0	0	
$+ 0$		Δ_j	0	0	0	0	0	0	
$= 6$									

$\therefore$ Z max. = 6 for $x_1 = 0$, $x_2 = 0$ and $x_3 = 4$ Ans.

Example 29. Max. $Z = 2x_1 + x_2$

Subject to

$$x_1 + 2x_2 \leq 10$$
$$x_1 + x_2 \leq 6$$
$$x_1 - x_2 \leq 2$$
$$x_1 - 2x_2 \leq 1$$
$$x_1, x_2 \geq 0$$

Equations obtained by introducing slack variables are

$$x_1 + 2x_2 + x_3 + 0.x_4 + 0.x_5 + 0.x_6 = 10$$
$$x_1 + x_2 + 0.x_3 + x_4 + 0.x_5 + 0.x_6 = 6$$
$$x_1 - x_2 + 0.x_3 + 0.x_4 + x_5 + 0.x_6 = 2$$
$$x_1 - 2x_2 + 0.x_3 + 0.x_4 + 0.x_5 + x_6 = 1$$

Taking $x_1 = 0, x_2 = 0,$

$x_3 = 10, x_4 = 6, x_5 = 2$ and $x_6 = 1$

objective function $Z = 2x_1 + x_2 + 0.x_3 + 0.x_4 + 0.x_5 + 0.x_6$

SIMPLEX TABLE 1

B	C_B	X_B	Y_1	Y_2	Y_3	Y_4	Y_5	Y_6	*Minimum ratio*
Y_3	0	10	1	2	1	0	0	0	10/1 = 10
Y_4	0	6	1	1	0	1	0	0	6/1 = 6
Y_5	0	2	1	–1	0	0	1	0	2/1 = 2
Y_6	0	1	$\boxed{1}$	–2	0	0	0	1	1/1 = 1 →
		x_j	0	0	10	6	2	1	
		c_j	2	1	0	0	0	0	
		Δ_j	2	1	0	0	0	0	
			↑ incoming vector					↓ outgoing vector	

SIMPLEX TABLE 2

B	C_B	X_B	Y_1	Y_2	Y_3	Y_4	Y_5	Y_6	*Minimum ratio* $= \frac{X_B}{Y_2}$
Y_3	0	9	0	4	1	0	0	–1	9/4
Y_4	0	5	0	3	0	1	0	–1	5/3
Y_5	0	1	0	$\boxed{1}$	0	0	1	–1	1/1 = 1 →
Y_1	2	1	1	–2	0	0	0	1	$\frac{1}{2}$ (neg.)
$Z = C_B X_B = 2$		x_j	1	0	9	5	1	0	
		c_j	2	1	0	0	0	0	
		Δ_j	0	5	0	0	0	–2	
				↑			↓		

$$\Delta_1 = c_1 - C_B Y_1 = 2 - (2 \times 1) = 0$$

$$\Delta_2 = c_2 - C_B Y_2 = 1 - \{2 \times (-2)\} = 5$$

$$\Delta_3 = c_3 - C_B Y_3 = 0 - 0 = 0$$

$$\Delta_4 = c_4 - C_B Y_4 = 0 - 0 = 0$$
$$\Delta_5 = c_5 - C_B Y_5 = 0 - 0 = 0$$
$$\Delta_6 = c_6 - C_B Y_6 = 0 - (2 \times 1) = -2$$

SIMPLEX TABLE 3

B	C_B	X_B	Y_1	Y_2	Y_3	Y_4	Y_5	Y_6	*Min. Ratio*
Y_3	0	5	0	0	1	0	– 4	3	5/3
Y_4	0	2	0	0	0	1	– 3	[2]	2/2 = 1 →
Y_2	1	1	0	1	0	0	1	– 1	1/–1 = –1
Y_1	2	3	1	0	0	0	2	– 1	3/–1 = –3
$Z = C_B X_B$ = 7	x_j	3	1	5	2	0	0	
	c_j	2	1	0	0	0	0	
	Δ_j	0	0	0	0	– 5	3	
						↓		↑	

$$\Delta_1 = 2 - (2 \times 1) = 0$$
$$\Delta_2 = 1 - (1 \times 1) = 0$$
$$\Delta_3 = 0 - 0 = 0$$
$$\Delta_4 = 0 - 0 = 0$$
$$\Delta_5 = 0 - (1 \times 1 + 2 \times 2) = -5$$
$$\Delta_5 = 0 - (1 \times -1 + 2 \times -1) = 3$$

SIMPLEX TABLE 4

B	C_B	X_B	Y_1	Y_2	Y_3	Y_4	Y_5	Y_6	*Min. Ratio*
Y_3	0	2	0	0	1	$-\frac{3}{2}$	$\frac{1}{2}$	0	
Y_6	0	1	0	0	0	$\frac{1}{2}$	$-\frac{3}{2}$	1	
Y_2	1	2	0	1	0	$\frac{1}{2}$	$-\frac{1}{2}$	0	
Y_1	2	4	1	0	0	$\frac{1}{2}$	$\frac{1}{2}$	0	
$Z = C_B X_B$ = 10	x_j	4	2	2	0	0	1	
	c_j	2	1	0	0	0	0	
	Δ_j	0	0	0	$-\frac{3}{2}$	$-\frac{1}{2}$	0	

All Δ_j's are either zero or negative

$\therefore$ Z max. = 10 for $x_1 = 4$ and $x_2 = 2$

Example 30. Solve by simplex method

Min. $Z = 7x_1 + x_2$

Subject to $3x_1 + 2x_2 \le -6$

$x_1 + 3x_2 \ge 15$

$x_1, x_2 \ge 0$

Solution: Given inequalities can be rearranged as

$$-3x_1 + 2x_2 \ge 6$$

$$x_1 + 2x_2 \ge 15$$

Introducing surplus variables x_3, x_4 we have

$$-3x_1 + 2x_2 - x_3 = 6$$

$$x_1 + 3x_2 - x_4 = 15$$

Introducing artificial variables x_5 and x_6 we get

$$-3x_1 + 2x_2 - x_3 + x_5 = 6 \quad \text{...(i)}$$

$$x_1 + 3x_2 - x_4 + x_6 = 15 \quad \text{...(ii)}$$

and objective function is

$$\text{Min. } Z = 7x_1 + x_2$$

So, objective function can be rearranged as

Max. $Z' = -(7x_1 + x_2)$

$= -7x_1 - x_2 + 0.x_3 + 0.x_4$ ...(iii)

Putting $x_1 = 0, x_2 = 0, x_3 = 0, x_4 = 0$ in (i) and (ii) we get

$$x_5 = 6 \text{ and } x_6 = 15$$

Then starting B.F.S. are

$$x_1 = 0, x_2 = 0, x_3 = 0, x_4 = 0$$

$$x_5 = 6 \text{ and } x_6 = 15.$$

Phase I

(To remove artificial variable x_5)

TABLE 1

B	X_B	Y_1	Y_2	Y_3	Y_4	A_1	A_2	
A_1	6	−3	2	−1	0	1	0	Here $x_5 \equiv A_1$
A_2	15	1	[3]	0	−1	0	1	$\rightarrow x_6 \equiv A_2$
			↑				↓	

(To remove artificial variable x_6)

TABLE 2

B	X_B	Y_1	Y_2	Y_3	Y_4	Y_5	
A_1	4	$\boxed{\frac{11}{3}}$	0	1	$\frac{-2}{3}$	—	→
Y_2	5	$\frac{1}{3}$	1	0	$-\frac{1}{3}$	0	
		↑				↓	

TABLE 3

B	X_B	Y_1	Y_2	Y_3	Y_4
Y_1	$\frac{12}{11}$	1	0	$\frac{3}{11}$	$\frac{-2}{11}$
Y_2	$\frac{51}{11}$	0	1	$\frac{-1}{11}$	$\frac{-3}{11}$

Phase II

SIMPLEX TABLE 1

B	C_B	X_B	Y_1	Y_2	Y_3	Y_4	Min. Ratio $= X_B/Y_3$
Y_1	–7	$\frac{12}{11}$	1	0	$\boxed{\frac{3}{11}}$	$-\frac{2}{11}$	$\frac{\frac{12}{11}}{\frac{3}{11}} = 4$ →
Y_2	–1	$\frac{51}{11}$	0	1	$-\frac{1}{11}$	$-\frac{3}{11}$	$\frac{\frac{51}{11}}{\frac{3}{1}} = \frac{51}{3} = 17$
$Z = C_B X_B$		x_j	$\frac{12}{11}$	$\frac{51}{11}$	0	0	
$= -7 \times \frac{12}{11}$		c_j	–7	–1	0	0	
$+ -1 \times \frac{51}{11}$							
$= -\frac{135}{11}$		Δ_j	0	0	$\frac{20}{11}$	–1	
			↓		↑		

B	C_B	X_B	Y_1	Y_2	Y_3	Y_4
Y_3	0	4	$\frac{11}{3}$	0	1	$-\frac{2}{3}$
Y_2	–1	5	$\frac{1}{3}$	1	0	– 3
$Z = C_B X_B$ $= 0 \times 4$		x_j	0	5	4	0
$+ (-1) \times 5$		c_j	–7	–1	0	0
$= -5$		Δ_j	$-\frac{20}{3}$	0	0	–3

$\therefore \quad Z' \text{ Max.} = -5$

$\therefore \quad Z \text{ Min.} = -(-5) = 5$

for $\quad x_1 = 0, x_2 = 5$ Ans.

Example 31. Min.$Z = -3x_1 + 2x_2$

Subject to $\quad x_1 - 4x_2 \leq -14$

$$-3x_1 + 2x_2 \geq 6$$

$$x_1, x_2 \geq 0$$

Solution: The constraints can be rearranged as

$$-x_1 + 4x_2 \geq 14$$

and $\quad -3x_1 + 2x_2 \geq 6$

Introducing, surplus variables x_3, x_4 artificial variables x_4 and x_5, we get

$$-x_1 + 4x_2 - x_3 + x_5 = 14 \quad \text{...(i)}$$

$$-3x_1 + 2x_2 - x_4 + x_6 = 6 \quad \text{...(ii)}$$

and objective function is

$$\text{Max. } Z' = -(\text{Min } Z)$$

$$= 3x_1 - 2x_2 + 0.x_3 + 0.x_4 + 0.x_5 + 0x_6 \quad \text{...(iii)}$$

Putting $x_1 = 0, x_2 = 0, x_3 = 0$
and $x_4 = 0$ in (i), (ii) and (iii) we get

$$x_5 = 14 \text{ and } x_6 = 6$$

So that starting B.F.S.

$$x_1 = 0, x_2 = 0, x_3 = 0, x_4 = 0, x_5 = 14$$

and $\quad x_6 = 6$

Phase I

(To remove artificial variable x_6)

TABLE 1

B	X_B	Y_1	Y_2	Y_3	Y_4	A_1	A_2	
A_1	14	–1	4	–1	0	1	0	
A_2	6	–3	[2]	0	–1	0	1	→
			↑				↓	

(To remove artificial variable x_5)

TABLE 2

B	X_B	Y_1	Y_2	Y_3	Y_4	A_1	
A_1	2	[5]	0	–1	2	1	→
Y_2	3	$-\frac{3}{2}$	1	0	$-\frac{1}{2}$	0	
		↑				↓	

TABLE 3

B	X_B	Y_1	Y_2	Y_3	Y_4
Y_1	$\frac{2}{5}$	1	0	$\frac{-1}{5}$	$\frac{2}{5}$
Y_2	$\frac{18}{5}$	0	1	$-\frac{3}{10}$	$\frac{1}{10}$

Phase II

Simplex Table Formation

SIMPLE TABLE

B	C_B	X_B	Y_1	Y_2	Y_3	Y_4	*Minimum Ratio*
Y_1	3	$\frac{2}{5}$	1	0	$-\frac{1}{5}$	$\frac{2}{5}$	
Y_2	–2	$\frac{18}{5}$	0	1	$-\frac{3}{10}$	$\frac{1}{10}$	
$Z' = C_B X_B$		x_j	$\frac{2}{5}$	$\frac{18}{5}$	0	0	
$= 3 \times \frac{2}{5} +$		c_j	3	–2	0	0	
$(-2) \times \frac{18}{5}$ $= -6$		Δ_j	0	0	0	–1	

$$Z' \text{ Max.} = -6$$

$\therefore$
$$Z \text{ min.} = -(-6) = 6$$

for $\left.\begin{matrix} x_1 = \dfrac{2}{5}, & x_2 = \dfrac{18}{5} \end{matrix}\right\}$ Ans.

Example 32. Max. $Z = 5x_1 + 3x_2$

Subject to
$$3x_1 + x_2 \le 11$$
$$x_1 + 2x_2 \le 7$$
$$x_1, \quad x_2 \ge 0$$

Solution: Introduce the slack variables x_3, x_4 to convert the inequalities into equations.

$\therefore$
$$3x_1 + x_2 + x_3 = 11$$
$$x_1 + 2x_2 + x_4 = 7$$

The equation can be rearranged as—

$$3x_1 + x_2 + x_3 + 0.x_4 = 11 \quad ...(i)$$
$$x_1 + 2x_2 + 0.x_3 + x_4 = 7 \quad ...(ii)$$

and objective function, can be written as—

Max $\quad Z = 5x_1 + 3x_2 + 0.x_3 + 0.x_4 \quad ...(iii)$

Putting $x_1 = 0 = x_2$ in (i) and (ii) we get $x_3 = 11$, $x_4 = 7$.

Starting Basic feasible solutions are—

$$x_1 = 0, \quad x_2 = 0, \quad x_3 = 11, \quad x_4 = 7$$

We construct the first simplex table as—

SIMPLEX TABLE 1

B	C_B	X_B	Y_1	Y_2	Y_3	Y_4	*Min. Ratio*
Y_3	0	11	$\boxed{3}$	1	1	0	$\frac{11}{3} \rightarrow$
Y_4	0	7	1	2	0	1	$\frac{7}{1} = 7$
$Z = C_B X_B$		x_j	0	0	11	7	
$= 0.11 + 0.7$		c_j	5	3	0	0	
$= 0$		Δ_j	5	3	0	0	
			$\uparrow$		$\downarrow$		

SIMPLEX TABLE 2

B	C_B	X_B	Y_1	Y_2	Y_3	Y_4	*Min. Ratio*
Y_1	5	11/3	1	1/3	1/3	0	11
Y_4	0	10/3	0	3	−1/3	1	2 →
$Z = C_B X_B$		x_j	11/3	0	0	10/3	
$\frac{5 \times 11}{3} + \frac{0 \times 10}{3}$		c_j	5	3	0	0	
$= \frac{55}{3}$		Δ_j	0	4/3	−5/3	0	
				↑		↓	

SIMPLEX TABLE 3

B	C_B	X_B	Y_1	Y_2	Y_3	Y_4	*Min. Ratio*
Y_1	5	3	1	0	2/5	−1/5	
Y_2	3	2	0	1	−1/5	3/5	
$Z = C_B X_B$		x_j	3	2	0	0	
= 5.3 + 3.2		c_j	5	3	0	0	
= 21		Δ_j	0	0	−7/5	−4/5	

$\therefore$ Z max. = 21 for $x_1 = 3, x_2 = 2$

Example 33. Max. $Z = 5x_1 + 3x_2$

Subject to $3x_1 + 5x_2 \le 15$

$5x_1 + 2x_2 \le 10$

$x_1, \quad x_2 \ge 0$

Solution: Adding the slack variables x_3 and x_4 to the in equalities to convert them into equations

$$3x_1 + 5x_2 + x_3 = 15$$

$$5x_1 + 2x_2 + x_4 - 10$$

Then eqns. can be written as

$$3x_1 + 5x_2 + x_3 + 0.x_4 = 15 \quad \text{......(ii)}$$

$$5x_1 + 2x_2 + 0.x_3 + x_4 = 10 \quad \text{......(iii)}$$

Putting $x_1 = 0 = x_2$ in (ii) and (iii) we get $x_3 = 15, x_4 = 10$. So the starting B.F.S. are

$$x_1 = 0, \quad x_2 = 0, \quad x_3 = 15, \quad x_4 = 10$$

SIMPLEX TABLE 1

B	C_B	X_B	Y_1	Y_2	Y_3	Y_4	*Min. Ratio*
Y_3	0	15	3	5	1	0	15/3 = 5
Y_4	0	10	[5]	2	0	1	10/5 = 2 →
$Z = C_B X_B$		x_j	0	0	15	10	
$= 0.15 + 0.10$		c_j	5	3	0	0	
$= 0$		Δ_j	5	3	0	0	
			↑			↓	

SIMPLEX TABLE 2

B	C_B	X_B	Y_1	Y_2	Y_3	Y_4	*Min. Ratio*
Y_3	0	9	0	$\boxed{\frac{19}{5}}$	1	$-\frac{3}{5}$	$\frac{9 \times 5}{19} = \frac{45}{19}$
Y_1	5	2	1	$\frac{2}{5}$	0	$\frac{1}{5}$	$\frac{2 \times 5}{2} = 5$
$Z = C_B X_B$		x_j	2	0	9	0	
$= 0.9 + 5.2$		c_j	5	3	0	0	
$= 10$		Δ_j	0	1	0	−1	
				↑	↓		

SIMPLEX TABLE 3

B	C_B	X_B	Y_1	Y_2	Y_3	Y_4	*Min. Ratio*
Y_2	3	$\frac{45}{19}$	0	1	$\frac{5}{19}$	$-\frac{3}{19}$	
Y_1	5	$\frac{20}{19}$	1	0	$-\frac{2}{19}$	$\frac{5}{19}$	
$Z = C_B X_B$		x_j	$\frac{20}{19}$	$\frac{45}{19}$	0	0	
$= \frac{3.45}{19} + \frac{5.20}{19}$		c_j	5	3	0	0	
$= \frac{235}{19}$		Δ_j	0	0	$\frac{-5}{19}$	$\frac{-16}{19}$	

$\therefore$ Max. $Z = \dfrac{235}{19}$

for $x_1 = \dfrac{20}{19}, \quad x_2 = \dfrac{45}{19}$ Ans.

Example 34. Max.$Z = 3x_1 - x_2$

Subject to
$$2x_1 + x_2 \leq 2$$
$$x_1 + 3x_2 \leq 3$$
$$x_2 \leq 4$$
$$x_1, x_2 \geq 0$$

Introduce slack variables x_3, x_4, x_5

$$2x_1 + x_2 + x_3 + 0.x_4 + 0.x_5 = 2 \quad \text{...(ii)}$$
$$x_1 + 3x_2 + 0.x_3 + x_4 + 0.x_5 = 0 \quad \text{...(iii)}$$
$$0.x_1 + x_2 + 0.x_3 + 0.x_4 + x_5 = 4 \quad \text{...(iv)}$$

Putting $x_1 = 0 = x_2$ in (ii), (iii) and (iv) we get

$$x_3 = 2, x_4 = 3, x_5 = 4$$

So the starting B.F.S. are

$$x_1 = 0, x_2 = 0, x_3 = 2, x_4 = 3, x_5 = 4$$

SIMPLEX TABLE 1

B	C_B	X_B	Y_1	Y_2	Y_3	Y_4	Y_5	*Min. Ratio*
Y_3	0	2	[2]	1	1	0	0	1 →
Y_4	0	3	1	3	0	1	0	3
Y_5	0	4	0		0	0	1	∞
$Z = C_B X_B$ = 0.2 + 0.3+ 9.4 = 0	x_j	0	0	2	3	4		
	c_j	3	−1	0	0	0		
	Δ_j	3	−1	0	0	0		
		↑		↓				

SIMPLEX TABLE 2

B	C_B	X_B	Y_1	Y_2	Y_3	Y_4	Y_5
Y_1	3	1	1	$\frac{1}{2}$	$\frac{1}{2}$	0	0
Y_4	0	2	0	$-\frac{5}{2}$	$-\frac{1}{2}$	1	0
Y_5	0	4	0	1	0	0	1
$Z = C_B Z_B$ = 3.1 + 0.2 + 0.4 = 3		x_j	1	0	0	2	4
		cj	3	−1	0	0	0
		Δ_j	0	$-\frac{5}{2}$	$-\frac{3}{2}$	0	0

∴ Max. $Z = 3, x_1 = 1, x_2 = 0$ Ans.

Example 35. Solve by simplex method

$$\text{Max. } Z = x_1 + x_2$$

Subect to $\quad -x_1 + 2x_2 \le 8$

$$3x_1 + x_2 \le 11;\ x_1, x_2 \ge 0$$

Solution: Add slack variables x_3 and x_4 to the constraints to convert them into equality. Then

$$-x_1 + 2x_2 + x_3 = 8$$

$$3x_1 + x_2 + x_4 = 11$$

These eqns. can be written as

$$-x_1 + 2x_2 + x_3 + 0.x_4 = 8$$

$$3x_1 + x_2 + 0x_3 + x_4 = 11$$

The objective function can be written as

$$Z = x_1 + x_2 + 0x_3 + 0x_4$$

Putting $\quad x_1 = 0, x_2 = 0$ starting basic Feasible

$$x_3 = 8 \text{ and } x_4 = 11$$

The first simplex table is constructed as follows

TABLE 1

B	C_B	X_B	Y_1	Y_2	Y_3	Y_4	*Minimum ratio* X_B/*incoming*
Y_3	0	8	–1	[2]	1	0	$\frac{8}{2} = 4$ →
Y_4	0	11	3	1	0	1	$\frac{11}{1} = 11$
$Z = C_B X_B = (0, 0)(8, 11)$		x_j c_j Δ_j	0 1 1	0 1 1	8 0 0	11 0 0	

↑ Incoming vector $\qquad$ ↓ Outgoing vector

$$\Delta_j = c_j - C_B Y_j$$

$$\Delta_1 = 1 - (0, 0)(-1, 3) = 1$$

Similarly $\quad \Delta_2 = 1, \quad \Delta_3 = 0, \quad \Delta_4 = 0$

Clearly $\quad \Delta_j \nleq 0$, for all the $Z = 0$ is not optimal solution

TABLE 2

B	C_B	X_B	Y_1	Y_2	Y_3	Y_4	*Min. Ratio*
Y_2	1	4	$-\frac{1}{2}$	1	$\frac{1}{2}$	0	– 8
Y_4	0	7	$\boxed{\frac{7}{2}}$	0	$-\frac{1}{2}$	1	2 →
$Z = C_B X_B = 4$		x_j	0	4	0	7	
		c_j	1	1	0	0	
		Δ_j	$\frac{3}{2}$	0	$\frac{-1}{2}$	0	

↑ Incoming vector (under Y_1)   ↓ Outgoing vector (under Y_4)

$$\Delta_j = c_j - C_B Y_j$$

$$\Delta_1 = 1 - (1, 0)\left(-\frac{1}{2}, \frac{7}{2}\right)$$

$$= 1 + \frac{1}{2} = \frac{3}{2}$$

Similarly, $\Delta_2 = 0, \Delta_3 = \frac{1}{2}, \Delta_4 = 0$

Clearly Δ_j all 0 for all j $\therefore$ $Z = 4$ is not optimal solution

TABLE 3

B	C_B	X_B	Y_1	Y_2	Y_3	Y_4	*Min. Ratio*
Y_2	1	5	0	1	$\frac{3}{7}$	$\frac{1}{7}$	
Y_1	1	2	1	0	$-\frac{1}{7}$	$\frac{2}{7}$	
$Z = C_B X_B = 7$		x_j	2	5	0	0	
		c_j	1	1	0	0	
		Δ_j	0	0	$-\frac{2}{7}$	$-\frac{3}{7}$	

Clearly $\Delta_j \leq 0$, for all j.

Z Max. $= 7$ for $x_1 = 2$ and $x_2 = 5$ Ans.

Example 36. Max.$Z = x_1$

Subject to
$$3x_1 + 2x_2 \leq 12$$
$$2x_1 + 3x_2 \leq 13$$
$$x_1, x_2 \geq 0$$

Solution: Add slack variables x_3 and x_4 the constraints to bring them into equally.

$$3x_1 + 2x_2 + x_3 = 12$$
$$2x_1 + 3x_2 + x_4 = 13$$

These equations can be written as

$$3x_1 + 2x_2 + x_3 + 0x_4 = 12$$
$$2x_1 + 3x_2 + 0x_3 + x_4 = 13$$

The objective function can be written as

$$Z = x_1 + 0x_2 + 0x_3 + 0x_4$$

The starting B.F.S. is obtained by putting $x_1 = x_2 = 0$ then

$$x_3 = 12 \text{ and } x_4 = 13$$

The 1st simplex table is constructed as follows

TABLE 1

B	C_B	X_B	Y_1	Y_2	Y_3	Y_4	*Min. Ratio*
Y_3	0	12	[3]	2	1	0	4 →
Y_4	0	13	2	3	0	1	$\frac{13}{2}$
$Z = C_B X_B$ = 0		x_j c_j Δ_j	0 1 1	0 0 0	12 0 0	13 0 0	

↑ Incoming vector ↓ Outgoing vector

$$\Delta_j = c_j - C_B Y_j$$
$$\Delta_1 = 1 - (0, 0)(3, 2)$$
$$= 1 - 0 = 1$$

Similarly $\Delta_2 = 0, \Delta_3 = 0, \Delta_4 = 0$

Clearly Δ_j 0, $\forall j$, Hence $Z = 0$ is not optimal solution.

TABLE 2

B	C_B	X_B	Y_1	Y_2	Y_3	Y_4	Minmum ratio
Y_1	1	4	1	$\frac{2}{3}$	$\frac{1}{3}$	0	
Y_4	0	5	0	$\frac{5}{3}$	$-\frac{2}{3}$	1	
$Z = C_B X_B$ $= 1 \times 4$ $= 4$		x_j c_j Δ_j	4 1 0	0 0 $\frac{-2}{3}$	0 0 $\frac{-1}{3}$	5 0 0	

$$\Delta_j = c_j - C_B Y_j$$

$$\Delta_1 = 1 - (1, 0)(1, 0) = 1 - 1 = 0$$

Similarly $\Delta_2 = \frac{-2}{3}$

$$\Delta_3 = \frac{-1}{3} \quad \Delta_4 = 0$$

Clearly $\Delta_j \le 0$ for $\forall_j$ Hence $Z = 4$ is optimal solution.

Z max. $= 4$ for $x_1 = 4, x_2 = 0$

Example 37. Max. $Z = x_2$

Subject to

$$3x_1 + 2x_2 \le 12$$

$$2x_1 + 3x_2 \le 13$$

$$x_1, x_2 \ge 0$$

Solution: Add slack variables x_3 and x_4 to the constraints to bring them into equality. Then,

$$3x_1 + 2x_2 + x_3 = 12$$

$$2x_1 + 3x_2 + x_4 = 13$$

These eqns can be written as

$$3x_1 + 2x_2 + x_3 + 0x_4 = 12$$

$$2x_1 + 3x_2 + 0x_3 + x_4 = 13$$

The objective function can be written as

$$Z = 0x_1 + x_2 + 0x_3 + 0x_4$$

Putting $x_1 = x_2 = 0$

we get $x_3 = 12$ and $x_4 = 13$.

TABLE 1

B	C_B	X_B	Y_1	Y_2	Y_3	Y_4	*Min. Ratio*
Y_3	0	12	3	2	1	0	$\frac{12}{2} = 6$
Y_4	0	13	2	$\boxed{3}$	0	1	$\frac{13}{2} \rightarrow$
$Z = C_B X_B$ $= 0$		x_j c_j Δ_j	0 0 0	0 1 1	12 0 0	13 0 0	

↑ Incoming vector ↓ Outgoing vector

$\boxed{3}$ is key element

$$\Delta_j = C_j - C_B Y_j$$

$$\Delta_1 = 0 - (0, 0)(3, 2) = 0$$

$$\Delta_2 = 0 - (0, 0)(2, 3) = 1$$

$$\Delta_3 = 0 - (0, 0)(1, 0) = 0$$

$$\Delta_4 = 0 - (0, 0)(0, 1) = 0$$

Clearly $\Delta_j \nleq 0$, then $Z = 0$ is not optimal solution.

TABLE 2

B	C_B	X_B	Y_1	Y_2	Y_3	Y_4	*Min. Ratio*
Y_3	0	$\frac{10}{3}$	$\frac{5}{3}$	0	1	$-\frac{2}{3}$	
Y_2	1	$\frac{13}{2}$	$\frac{2}{3}$	1	0	$\frac{1}{3}$	
$Z = C_B X_B$		x_j	0	$\frac{13}{3}$	$\frac{10}{3}$	0	
$= \frac{13}{3}$		c_j	0	1	0	0	
		Δ_j	$-\frac{2}{3}$	0	0	$-\frac{1}{3}$	

$$\Delta_j = C_j - C_B Y_j$$

$$\Delta_1 = 0 - (0, 1)\left(\frac{5}{2}, \frac{2}{3}\right) = \frac{2}{3}$$

Similarly $\Delta_2 = 0, \Delta_3 = 0, \Delta_4 = -\frac{1}{3}$

Clearly $\Delta_j \leq 0$, So $Z = \frac{13}{3}$ is optimal sol.

$$Z \max = \frac{13}{3} \quad \text{for} \quad x_1 = 0, \quad x_2 = \frac{13}{3}$$

Example 38. Max. $Z = 13x_1 + 30x_2$

Subject to

$$x_1 - x_2 \leq 3$$

$$3x_1 + x_2 \leq 11$$

$$x_1 + 2x_2 \leq 7 \quad x_1, x_2 \geq 0$$

Solution: Add slack variables x_3, x_4 and x_5 to the constraints to convert them into equations.

$$x_1 - x_2 + x_3 = 3$$

$$3x_1 + x_2 + x_4 = 11$$

$$x_1 + 2x_2 + x_5 = 7$$

These equations are

$$x_1 - x_2 + x_3 + 0x_4 + 0x_5 = 3$$

$$3x_1 + x_2 + 0x_3 + x_4 + 0x_5 = 11$$

$$x_1 + 2x_2 + 0x_3 + 0x_4 + x_5 = 7$$

The objective function can be written as

$$Z = 13x_1 + 30x_2 + 0x_3 + 0x_4 + 0x_5$$

Putting $x_1 = 0, x_2 = 0$

the starting basic feasible solution is

$$x_3 = 3, \quad x_4 = 11, \quad x_5 = 7$$

The 1st simplex table is constructed as follows:

TABLE 1

B	C_B	X_B	Y_1	Y_2	Y_3	Y_4	Y_5	*Min. Ratio*
Y_3	0	3	1	−1	1	0	0	$\frac{3}{-1}$
Y_4	0	11	3	1	0	1	0	$\frac{11}{1} = 11$
Y_5	0	7	1	[2]	0	0	1	$\frac{7}{2} = 3.5 \rightarrow$
$Z = C_B X_B$		x_j	0	0	3	11	7	
$= 0$		C_j	13	30	0	0	0	
		Δ_j	13	30	0	0	0	
				↑ Incoming vector			↓ Outgoing vector	

TABLE 2

B	C_B	X_B	Y_1	Y_2	Y_3	Y_4	Y_5	*Min. Ratio*
Y_3	0	$\frac{13}{2}$	$\frac{3}{2}$	0	1	0	$\frac{1}{2}$	
Y_4	0	$\frac{15}{2}$	$\frac{5}{2}$	0	0	1	$-\frac{1}{2}$	
Y_2	30	$\frac{7}{2}$	$\frac{1}{2}$	1	0	0	$\frac{1}{2}$	
$Z = C_B X_B$		x_j	0	$\frac{7}{2}$	$\frac{13}{2}$	$\frac{15}{2}$	0	
$= 105$		c_j	13	30	0	0	0	
		Δ_j	−2	0	0	0	−15	

$$\Delta_j = C_j - C_B Y_j$$

$$\Delta_1 = 13 - (0, 0, 30)\left(\frac{3}{2}, \frac{5}{2}, \frac{1}{2}\right)$$

$$= 13 - 15 = -2$$

Similarly $\Delta_2 = 0$, $\Delta_3 = 0$, $\Delta_4 = 0$, $\Delta_5 = -15$

Clearly $\Delta_j \leq 0$ for $\forall j$. Hence $Z = 105$ is optimal.

Solution: Z max $= 105$ for $x_1 = 0$, $x_2 = \frac{7}{2}$ Ans.

Example 39. Max. $Z = 7x_1 + 11x_2$

$$3x_1 + x_2 \leq 11$$

$$x_1 + 2x_2 \leq 7$$

$$x_1 - x_2 \leq 3 \quad : \quad x_1, x_2 \geq 0$$

Solution: Add slack variables x_3, x_4 and x_5 to the constraint.

$$3x_1 + x_2 + x_3 = 11$$

$$x_1 + 2x_2 + x_4 = 7$$

$$x_1 - x_2 + x_5 = 3$$

These equations are as

$$3x_1 + x_2 + x_3 + 0x_4 + 0x_5 = 11$$

$$x_1 + 2x_2 + 0x_3 + x_4 + 0x_5 = 7$$

$$x_1 - x_2 + 0x_3 + 0x_4 + x_5 = 3$$

Putting $x_1 = x_2 = 0$, the starting B.F.S. $x_3 = 11, x_4 = 7, x_5 = 3$

The objective function can be written as,

$$Z = 7x_1 + 11x_2 + 0x_3 + 0x_4 + 0x_5$$

The first simplex table is constructed as follows.

TABLE 1

B	C_B	X_B	Y_1	Y_2	Y_3	Y_4	Y_5	*Min. Ratio*
Y_3	0	11	3	1	1	0	0	$\frac{11}{1} = 11$
Y_4	0	7	1	$\boxed{2}$	0	1	0	$\frac{7}{2} \rightarrow$
Y_5	0	3	1	-1	0	0	1	$\frac{3}{-1}$ rejected
$Z = C_B X_B$		x_j	0	0	11	7	3	
$= 0$		c_j	7	11	0	0	0	
		Δ_j	7	11	0	0	0	
				↑ Incoming vector		↓ Outgoing vector		

$\boxed{2}$ is key element

$$\Delta_j = c_j - C_B Y_j$$

$$\Delta_1 = 7 - (0, 0, 0)\,(3, 1, 1) = 7 - 0 = 7$$

Similarly $\Delta_2 = 11, \Delta_3 = 0, \Delta_4 = 0, \Delta_5 = 0$

Clearly, $\Delta_j \nless 0, \forall_j$. Hence $Z = 0$ is not optimal solution.

TABLE 2

B	C_B	X_B	Y_1	Y_2	Y_3	Y_4	Y_5	*Min. Ratio*
Y_3	0	$\frac{15}{2}$	$\boxed{\frac{5}{2}}$	0	1	$-\frac{1}{2}$	0	3 →
Y_2	11	$\frac{7}{2}$	$\frac{1}{2}$	1	0	$\frac{1}{2}$	0	7
Y_5	0	$\frac{13}{2}$	$\frac{3}{2}$	0	0	$+\frac{1}{2}$	1	$\frac{13}{3}$
$Z = C_B X_B$		x_j	0	$\frac{7}{2}$	$\frac{15}{2}$	0	$\frac{13}{2}$	
$= \frac{77}{2}$		c_j	7	11	0	0	0	
		Δ_j	$\frac{3}{2}$	0	0	$-\frac{11}{2}$	0	

↑ Incoming vector (under Y_1) ↓ Outgoing vector (under Y_3)

TABLE 3

B	C_B	X_B	Y_1	Y_2	Y_3	Y_4	Y_5	*Min. Ratio*
Y_1	7	3	1	0	$\frac{2}{5}$	$-\frac{1}{5}$	0	
Y_2	11	2	0	1	$-\frac{1}{5}$	$\frac{2}{5}$	0	
Y_5	0	$\frac{7}{2}$	0	0	$-\frac{3}{5}$	$\frac{4}{5}$	1	
$Z = C_B X_B$		x_j	3	2	0	0	$\frac{7}{2}$	
$= 43$		c_j	7	11	0	0	0	
		Δ_j	0	0	$-\frac{3}{5}$	$-\frac{26}{5}$	0	

$$\Delta_j = c_j - C_B Y_j$$

$$\Delta_1 = 7 - (7, 11, 0)(1, 0, 0) = 7 - 7 = 0$$

Similarly $\Delta_2 = 0, \quad \Delta_3 = \frac{-3}{5}, \quad \Delta_4 = \frac{-26}{5}, \quad \Delta_5 = 0$

Clearly $\Delta_j \le 0 \; \forall \, j$ hence $Z = 43$ is optimal solution.

Z max = 43 for $x_1 = 3$ and $x_2 = 2$

Example 40. Max. $Z = x_2$

Subject to

$$-3x_1 + 2x_2 \le 6$$

$$2x_1 + x_2 \le 14$$

$$x_1 - x_2 \le 4; x_1, x_2 \ge 0$$

Solution: Add slack variables x_3, x_4 and x_5 to the constraints to convert them into equation.

$$-3x_1 + 2x_2 + x_3 = 6$$

$$2x_1 + x_2 + x_4 = 14$$

$$x_1 - x_2 + x_5 = 4$$

These equations are

$$-3x_1 + 2x_2 + x_3 + 0x_4 + 0x_5 = 6$$

$$2x_1 + x_2 + 0x_3 + 0x_4 + 0x_5 = 14$$

$$x_1 - x_2 + 0x_3 + 0x_4 + x_5 = 4$$

The objective function can be written as

$$Z = 0x_1 + x_2 + 0x_3 + 0x_4 + 0x_5$$

The starting basic feasible solution can be obtained by putting decision variables equal to zero, then,

$$x_3 = 6, x_4 = 14, x_5 = 4$$

TABLE 1

B	C_B	X_B	Y_1	Y_2	Y_3	Y_4	Y_5	*Min. Ratio*
Y_3	0	6	–3	$\boxed{2}$	1	0	0	$3 \rightarrow$
Y_4	0	14	2	1	0	1	0	14
Y_5	0	4	1	–1	0	0	1	$\frac{4}{-1}$
$Z = C_B X_B$		x_j	0	0	6	14	4	
$= 0$		c_j	0	1	0	0	0	
		Δ_j	0	1	0	0	0	
				↑ Incoming vector		↓ Outgoing vector		

$\boxed{2}$ is key element.

$$\Delta_j = c_j - C_B Y_j$$

$$\Delta_1 = 0 - (0, 0, 0)(-3, 2, 1) = 0 - 0 = 0$$

Similarly $\Delta_2 = 1, \Delta_3 = 0, \Delta_4 = 0$ and $\Delta_5 = 0$

Clearly $\Delta_j \leq \forall j, Z = 0$ is not optimal solution.

TABLE 2

B	C_B	X_B	Y_1	Y_2	Y_3	Y_4	Y_5	*Min. Ratio*
Y_2	1	3	$\frac{3}{-2}$	1	$\frac{1}{2}$	0	0	–2
Y_4	0	11	$\boxed{\frac{7}{2}}$	0	$-\frac{1}{2}$	1	0	$\frac{22}{7} \rightarrow$
Y_5	3	7	$-\frac{1}{2}$	0	$\frac{1}{2}$	0	1	–14
$Z = C_B X_B$		x_j	3	3	0	11	7	
$= 3$		c_j	0	1	0	0	0	
		Δ_j	$\frac{3}{2}$	0	$-\frac{1}{2}$	0	0	
			↑ Incoming vector			↓ Outgoing vector		

$\boxed{\frac{7}{2}}$ is key element

TABLE 3

B	C_B	X_B	Y_1	Y_2	Y_3	Y_4	Y_5	Min. Ratio
Y_2	1	$\frac{54}{7}$	0	1	$+\frac{2}{7}$	$\frac{3}{7}$	0	
Y_1	0	$\frac{22}{7}$	1	0	$-\frac{1}{7}$	$\frac{2}{7}$	0	
Y_5	0	$\frac{60}{7}$	0	0	$\frac{3}{7}$	$\frac{1}{7}$	1	
$Z = C_B X_B$		x_j	$\frac{22}{7}$	$\frac{54}{7}$	0	0	$\frac{60}{7}$	
$= \frac{54}{7}$		c_j	0	1	0	0	0	
		Δ_j	0	0	$-\frac{2}{7}$	$-\frac{3}{7}$	0	

$$\Delta_j = c_j - C_B Y_j$$

$$\Delta_1 = 0 - (1, 0, 0)(0, 1, 0) = 0 - 0 = 0$$

$$\Delta_2 = 1 - (1, 0, 0)(1, 0, 0) = 1 - 1 = 0$$

Similarly $\Delta_3 = -\frac{2}{7}, \quad \Delta_4 = -\frac{3}{7}$ and $\Delta_5 = 0$

Clearly $\Delta_j \le 0$ for $\forall\, j$. Hence $Z = \frac{54}{7}$ is optimal solution.

$$Z \text{ max.} = \frac{54}{7} \text{ for } x_1 = \frac{22}{7}, x_2 = \frac{54}{7}$$

Example 41. Max. $Z = x_2$

Subject to

$$-3x_1 + 2x_2 \le 6$$
$$2x_1 + x_2 \le 14$$
$$x_1 - x_2 \le 4$$
$$2x_1 + 3x_2 \le 22; \quad x_1, x_2 \ge 0$$

Solution: Adding slack variables x_3, x_4, x_5, x_6 to constraints. We get

$$-3x_1 + 2x_2 + x_3 = 6$$
$$2x_1 + x_2 + x_4 = 14$$
$$x_1 - x_2 + x_5 = 4$$
$$2x_1 + 3x_2 + x_6 = 22$$

These can be written as

$$-3x_1 + 2x_2 + x_3 + 0x_4 + 0x_5 + 0x_6 = 6$$
$$2x_1 + x_2 + 0x_3 + x_4 + 0x_5 + 0x_6 = 14$$
$$x_1 - x_2 + 0x_3 + 0x_4 + x_5 + 0x_6 = 4$$
$$2x_1 + 3x_2 + 0x_3 + 0x_4 + 0x_5 + x_6 = 22$$

The objective function can be written as

$$Z = 0x_1 + x_2 + 0x_3 + 0x_4 + 0x_5 + 0x_6$$

Putting $x_1 = x_2 = 0$, the starting B.F.S.

$$x_3 = 6, x_4 = 14, x_5 = 4, x_6 = 22$$

The 1st simplex table is constructed as follows.

TABLE 1

B	C_B	X_B	Y_1	Y_2	Y_3	Y_4	Y_5	Y_6	*Min. Ratio*
Y_3	0	6	–3	[2]	1	0	0	0	$\frac{6}{2} = 3 \rightarrow$
Y_4	0	14	2	1	0	1	0	0	$\frac{14}{1} = 14$
Y_5	0	4	1	–1	0	0	1	0	$\frac{4}{-1}$
Y_6	0	22	2	3	0	0	0	1	$\frac{22}{3}$
$Z = C_B X_B$		x_j	0	0	6	14	4	22	
$= 0$		c_j	0	1	0	0	0	0	
		Δ_j	0	1	0	0	0	0	
				↑ Incoming vector		↓ Outgoing vector			

$$\Delta_j = c_j - C_B Y_j$$
$$\Delta_1 = 0 - (0, 0, 0, 0)(-3, 2, 1, 2)$$
$$= 0 - 0 = 0$$

Similarly, $\Delta_2 = 1, \Delta_3 = 0, \Delta_4 = 0, \Delta_5 = 0, \Delta_6 = 0$

[2] is key element

Clearly $\Delta_j \nleq 0; \forall j$. Hence $Z = 0$ is not optimal solution.

TABLE 2

B	C_B	X_B	Y_1	Y_2	Y_3	Y_4	Y_5	Y_6	*Min. Ratio*
Y_2	1	3	$-\frac{3}{2}$	1	$\frac{1}{2}$	0	0	0	-2
Y_4	0	11	$\frac{7}{2}$	0	$-\frac{1}{2}$	1	0	0	$\frac{22}{7}$
Y_5	0	7	$-\frac{1}{2}$	0	$-\frac{1}{2}$	0	1	0	-14
Y_6	0	13	$\boxed{\frac{13}{2}}$	0	$-\frac{3}{2}$ = 2 →	0	0	1	$\frac{13}{13} \times 2$
$Z = C_B X_B$		x_j	0	3	0	11	7	13	
$= 3$		c_j	6	1	0	0	0	0	
		Δ_j	$\frac{3}{2}$	0	$-\frac{1}{2}$	0	0	0	
			↑ Incoming vector					↓ Outgoing vector	

$\boxed{\frac{13}{2}}$ is key element

$Z = 3$ is not optimal solution as $\Delta_j \nless 0, \forall_j$

TABLE 3

B	C_B	X_B	Y_1	Y_2	Y_3	Y_4	Y_5	Y_6	*Min. Ratio*
Y_2	1	6	0	1	$\frac{2}{13}$	0	0	$+\frac{3}{13}$	
Y_4	0	4	0	0	$\frac{4}{13}$	1	0	$-\frac{7}{13}$	
Y_5	0	8	0	0	$-\frac{8}{13}$	0	1	$\frac{1}{3}$	
Y_1	0	2	1	0	$-\frac{3}{13}$	0	0	$\frac{3}{13}$	
$Z = C_B X_B$		x_j	2	6	0	4	8	0	
$= 6$		c_j	0	1	0	0	0	0	
		Δ_j	0	0	$-\frac{2}{13}$	0	0	$-\frac{3}{13}$	

Clearly $\Delta_j \le 0$ hence $Z = 6$ is optimal solution.

$$Z \max = 6 \quad \text{for} \quad x_1 = 2, x_2 = 6$$

Example 42. Max. $Z = x_1$ when the constraints are

$$3x_1 + x_2 \le 11$$
$$x_1 + 2x_2 \le 7$$
$$x_1 - x_2 \le 3$$
$$x_1, x_2 \ge 0$$

Solution: Adding slack variables x_3, x_4 and x_5 to the constraints We get

$$3x_1 + x_2 + x_3 = 11$$
$$x_1 + 2x_2 + x_4 = 7$$
$$x_1 - x_2 + x_5 = 3$$

These eqns. can be written as

$$3x_1 + x_2 + x_3 + 0x_4 + 0x_5 = 11$$
$$x_1 + 2x_2 + 0x_3 + x_4 + 0x_5 = 7$$
$$x_1 - x_2 + 0x_3 + 0x_4 + x_5 = 3$$

The objective function can be written as

$$Z = x_1 + 0x_2 + 0x_3 + 0x_4 + 0x_5$$

Putting $x_1 = x_2 = 0$, the starting B.F.S.

$$x_3 = 11, x_4 = 7, x_5 = 3$$

TABLE 1

B	C_B	X_B	Y_1	Y_2	Y_3	Y_4	Y_5	*Min. Ratio*
Y_3	0	11	3	1	1	0	0	$\frac{11}{3}$
Y_4	0	7	1	2	0	1	0	$\frac{7}{1}$
Y_5	0	3	$\boxed{1}$	−1	0	0	1	$\frac{3}{1} \rightarrow$
$Z = C_B X_B$		x_j	0	0	11	7	3	
= 0		c_j	1	0	0	0	0	
		Δ_j	1 ↑	0	0	0	0 ↓	
			Incoming vector				Outgoing vector	

$\boxed{1}$ is key element.

TABLE 2

B	C_B	X_B	Y_1	Y_2	Y_3	Y_4	Y_5	*Min. Ratio*
Y_3	0	2	0	$\boxed{4}$	1	0	-3	$\frac{1}{3} \rightarrow$
Y_4	0	4	0	3	0	1	-1	$\frac{4}{3}$
Y_1	1	3	1	-1	0	0	1	$\frac{3}{-1}$
$Z = C_B X_B$		x_j	3	0	2	4	0	
$= 3$		c_j	1	0	0	0	0	
		Δ_j	0	1	0	0	-1	
				↑ Incoming vector		↓ Outgoing vector		

$\boxed{4}$ is key element.

Clearly $\Delta_j \nleq 0, \forall j$ Hence $Z = 3$ is not optimal solution.

TABLE 3

B	C_B	X_B	Y_1	Y_2	Y_3	Y_4	Y_5	*Min. Ratio*
Y_2	0	$\frac{1}{2}$	0	1	$\frac{1}{4}$	0	$-\frac{3}{4}$	
Y_4	0	$\frac{5}{2}$	0	0	$-\frac{3}{4}$	1	$+\frac{5}{4}$	
Y_1	1	$\frac{7}{2}$	1	0	$\frac{1}{4}$	0	$\frac{1}{4}$	
$Z = C_B X_B$		x_j	$\frac{7}{2}$	$\frac{1}{2}$	0	$\frac{5}{2}$	0	
$(0, 0, 1)$		c_j	1	0	0	0	0	
$\left(\frac{1}{3}, 3, \frac{7}{3}\right)$		Δ_j	0	0	$-\frac{1}{4}$	0	$-\frac{1}{4}$	

Clearly $\Delta_j \leq 0$, hence $Z = \frac{7}{2}$ is optimal solution

$$Z \max = \frac{7}{2} \quad \text{for} \quad x_1 = \frac{7}{2}, \quad x_2 = \frac{1}{2}$$

Example 43. Max. $Z = x_2$

When the constraints are

$$\begin{aligned} 3x_1 + x_2 &\leq 11 \\ x_1 + 2x_2 &\leq 7 \\ x_1 - x_2 &\leq 3 \\ x_1, x_2 &\geq 0 \end{aligned}$$

Solution: Add slack variables x_3, x_4, x_5 to the constraints then

$$\begin{aligned} 3x_1 + x_2 + x_3 &= 11 \\ x_1 + 2x_2 + x_4 &= 7 \\ x_1 - x_2 + x_5 &= 3 \end{aligned}$$

These equations are

$$3x_1 + x_2 + x_3 + 0x_4 + 0x_5 = 11$$

$$x_1 + 2x_2 + 0x_3 + x_4 + 0x_5 = 7$$

$$x_1 - x_2 + 0x_3 + 0x_4 + 0x_5 = 3$$

The objective function can be written as

$$Z = 0x_1 + x_2 + 0x_2 + 0x_4 + 0x_5$$

Putting $x_1 = x_2 = 0$, the starting B.F.S.

$$x_3 = 11, x_4 = 7, x_5 = 3$$

TABLE 1

B	C_B	X_B	Y_1	Y_2	Y_3	Y_4	Y_5	*Min. Ratio*
Y_3	0	11	3	1	1	0	0	$\frac{11}{1} = 11$
Y_4	0	7	1	$\boxed{2}$	0	1	0	$\frac{7}{2} \rightarrow$
Y_5	3	3	1	-1	0	0	1	$\frac{3}{-1}$
$Z = C_B X_B$		x_j	0	0	11	7	3	
(0, 0, 0)		c_j	0	1	0	0	0	
(11, 7, 3) = 0		Δ_j	0	1	0	0	0	
				$\uparrow$		$\downarrow$		
				Incoming vector		Outgoing vector		

[2] is key element.

TABLE 2

B	C_B	X_B	Y_1	Y_2	Y_3	Y_4	Y_5	*Min. Ratio*
Y_3	0	$\frac{15}{2}$	$\frac{5}{2}$	0	0	$-\frac{1}{2}$	0	
Y_2	1	$\frac{7}{2}$	$\frac{1}{2}$	1	0	$\frac{1}{2}$	0	
Y_5	0	$\frac{13}{2}$	$\frac{3}{2}$	0	0	$\frac{1}{2}$	1	
$Z = C_B X_B$		x_j	0	$\frac{7}{2}$	$\frac{15}{2}$	0	$\frac{13}{2}$	
$= (0, 1, 0) \left(\frac{15}{2}, \frac{7}{2}, \frac{13}{2}\right)$		c_j	0	1	0	0	0	
$= \frac{7}{2}$		Δ_j	$-\frac{1}{2}$	0	0	$-\frac{1}{2}$	0	

Clearly $\Delta_j \leq 0$ for all j Hence $Z = \frac{7}{2}$ is optimal solution.

$$Z \max = \frac{7}{2} \text{ for } x_1 = 0, x_2 = \frac{7}{2} \qquad \text{Ans.}$$

EXERCISE

Solve by Simplex Method—

1. Max. $Z = 9x_1 + 7x_2$
   Subject to
   $x_1 + 2x_2 \leq 7$
   $x_1 - x_2 \leq 4$
   $x_1, x_2 \geq 0$
2. Max. $Z = 5x_1 + 3x_2$
   Subject to
   $3x_1 + x_2 \leq 11$
   $x_1 + 2x_2 \leq 7$
   $x_1, x_2 \geq 0$
3. Max. $Z = x_1 + 7x_2$
   $-x_1 + 2x_2 \leq 8$
   $x_1 - x_2 \leq 4$
   $x_1, x_2 \geq 0$

4. Max. $Z = x_1 + 3x_2$
Subject to
$3x_1 + 5x_2 \le 15$
$5x_1 + 2x_2 \le 10$
$x_1, x_2 \ge 0$

[Meerut '73]

5. Max. $Z = x_1 - x_2 + 3x_3$
Subject to
$x_1 + x_2 + x_3 \le 10$
$2x_1 - x_3 \le 2$
$2x_1 - 2x_2 + 3x_3 \le 0$
$x_1, \quad x_2, \quad x_3 \ge 0$

[Delhi '58, Meerut '78]

6. Max. $Z = 10x_1 + x_2 + 2x_3$
Subject to
$x_1 + x_2 - 2x_3 \le 10$
$4x_1 + x_2 + x_3 \le 20$
$x_1, \quad x_2, \quad x_3 \ge 0$

[Calcutta '77]

7. Min. $Z = 7x_1 - x_2$
Subject to
$3x_1 - 2x_2 \le -6$
$x_1 + 3x_2 \ge 15$
$x_1, x_2 \ge 0$

8. Min. $Z = -3x_1 + 2x_2$
Subject to
$x_1 - 4x_2 \le - 14$
$- 3x_1 + 2x_2 \le 6$
$x_1, x_2 \ge 0$

9. Max. $Z = 3x_1 - x_2$
Subject to
$2x_1 + x_2 \le 2$
$x_1 + 3x_2 \le 3$
$x_3 \le 4$
$x_1, x_2 \ge 0$

10. Max. $Z = 2x_1 + 4x_2 + x_3$
Subject to
$x_1 + 2x_2 \le 4$
$2x_1 + x_2 \le 3$
$x_2 + 4x_3 \le 3$
$x_1, \quad x_2, \quad x_3 \ge 0$

11. Max. $Z = 3x_1 + 2x_2$
Subject to
$2x_1 + x_2 \le 10$
$x_1 + 3x_2 \le 6$
$x_1, x_2 \ge 0$

11. (a) Max. $Z = 7x_1 + 11x_2$
Subject to
$3x_1 + x_2 \le 11$
$x_1 + 2x_2 \le 7$
$x_1 - x_2 \le 3$
$x_1, \quad x_2 \ge 0$

12. Max. $Z = 2x_1 + x_2$

Subject to
$x_1 + 2x_2 \le 10$
$x_1 + x_2 \le 6$
$x_1 - x_2 \le 2$
$x_1 - 2x_2 \le 1$
$x_1, x_2 \ge 0$

13. Write short notes on the simplex method.

[Meerut O.R. '72, '74]

or

Give computational procedure for simplex method in L.P.P.

[Meerut O.R. 74(S), 75(S), 77]

Use two phase simplex method to solve the following L.P.P.

14. Max. $Z = 3x_1 - x_2$

Subject to
$2x_1 + x_2 \ge 2$
$x_1 + 3x_2 \le 2$
$x_2 \le 4$
$x_1, x_2 \ge 0$

15. Max. $Z = 5x_1 + 8x_2$

Subject to
$3x_1 + 2x_2 \ge 3$
$x_1 + 4x_2 \ge 4$
$x_1 + x_2 \le 5$
$x_1 \ge 0, x_2 \ge 0$

Use Big M-method to solve the following L.P.P.

16. Min. $Z = 2x_1 + 9x_2 + x_3$

Subject to
$x_1 + 4x_2 + 2x_3 \ge 5$
$3x_1 + x_2 + 2x_3 \ge 4$
$x_1, \quad x_2, \quad x_3 \ge 0$

17. Max. $Z = 5x - 2y + 3z$

Subject to
$2x + 2y - z \ge 2$
$3x - 4y \le 3$
$y + 3z \le 5$
$x, \quad y, \quad z \ge 0$

ANSWERS

1. $Z_{max} = 52, x_1 = 5, x_2 = 1$
2. $Z_{max} = 21, x_1 = 3, x_2 = 2$
3. $Z_{max} = 100, x_1 = 16, x_2 = 12$
4. $Z_{max} = \frac{235}{19}, \quad x_1 = \frac{20}{19}, \quad x_2 = \frac{45}{19}$
5. $Z_{max} = 6, x_1 = 0, x_2 = 6, x_3 = 4$
6. $Z_{max} = 50, x_1 = 5, x_2 = 0, x_3 = 0$

7. $Z_{max} = 5, x_1 = 0, x_2 = 5$

8. $Z_{min} = 6, x_1 = \frac{2}{5}, x_2 = \frac{18}{5}$

9. $Z_{max} = 3, x_4 = 1, x_2 = 0$

10. $Z_{max} = \frac{25}{3}, x_1 = \frac{2}{3}, x_2 = \frac{5}{3}, x_3 = \frac{1}{5}$

11. $Z_{max} = \frac{76}{5}, x_1 = \frac{24}{5}, x_2 = \frac{2}{5}$

11.(a) $Z_{max} = 43, \quad x_1 = 3, \quad x_2 = 2$

12. $Z_{max} = 10, x_1 = 4, x_2 = 2$

14. $Z_{max} = 40, x_1 = 0, x_2 = 5$

15. $Z_{max} = \frac{5}{2}, x_1 = 0, x_2 = 0, x_3 = \frac{5}{2}$

16. $Z_{max} = \frac{101}{9}, x = \frac{13}{9}, y = \frac{1}{3}, z = \frac{14}{9}$

3

Convex Sets

In this chapter we shall discuss the basic properties of convex sets in n-dimensional Euclidean space R^n. This concept of convexity of sets depends upon the concepts of *neighbourhood* briefly denoted by *nhd* and *line-segment*. So firstly some useful definitions will be discussed.

3.1. EUCLIDEAN SPACE R^n

We know that the position vector of $P(x_1, x_2)$ is

$\vec{r} = x_1\vec{i} + x_2\vec{j}$ where

$\vec{i}, \vec{j}$ are unit vectors along Ox_1 and Ox_2, respectively.

We write $\vec{r} = x_1\vec{i} + x_2\vec{j}$ as

$$\vec{r} = (x_1, x_2).$$

Thus $\vec{r} = (x_1, x_2)$ represents a point in two dimensional space R^2.

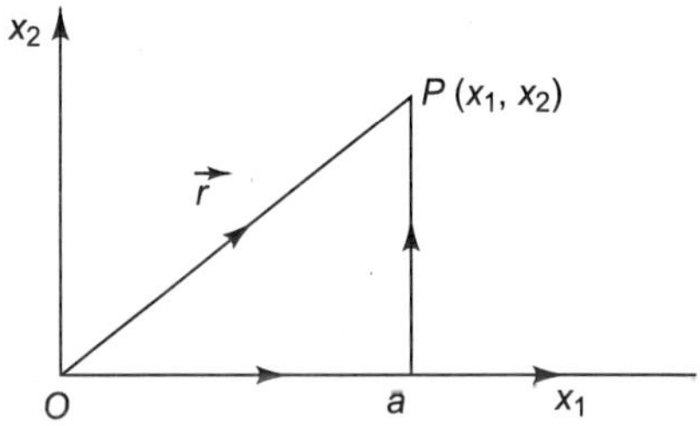

Fig. 3.1

The position vector of $P(x_1, x_2, x_3)$ is

$\vec{r} = x_1\vec{i} + x_2\vec{j} + x_3\vec{k}$

$\vec{i}, \vec{j}, \vec{k}$ are unit vectors along Ox_1, Ox_2 and Ox_3 respectively.

Now, $\vec{r} = x_1\vec{i} + x_2\vec{j} + x_3\vec{k}$ can be written as $\vec{r} = (x_1, x_2, x_3)$ which represents a point in three dimensional space R^3.

Proceeding in this way, we see that

$$\vec{r} = x_1\vec{e_1} + x_2\vec{e_2} + \ldots + x_n\vec{e_n} = (x_1, x_2, \ldots, x_n)$$

represents a point in n-dimensional space R^n, when $\vec{e}_1, \vec{e}_2, \vec{e}_n$ are unit vector.

We are in position to define the n-dimensional space R^n as below. The set of all ordered n-tuples of scalars, that is, the set in which every element (or point) is represented by means of n co-ordinates, is called the n-dimensional space R^n.

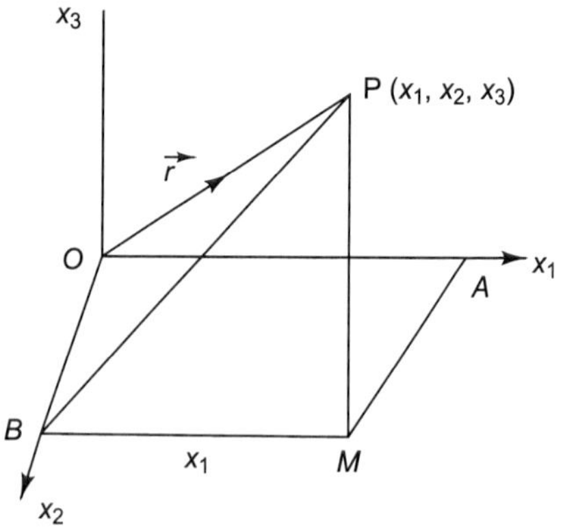

Fig. 3.2

In $\vec{r} = (x_1, x_2, ..., x_n)$, $x_1, x_2, x_3, ..., x_n$ are n-tuples scalars and unit vectors are generally denoted by $\vec{e}_1, \vec{e}_2, ... \vec{e}_n$.

The space R^n *is called n-dimenstional Euclidean space*. In our discussion the vector $a \in R^n$ will represents a point in R^n when vector

$$\vec{a} = (a_1, a_2, ..., a_n).$$

3.2. POINT SET

Points set is the set whose elements are points or vector in R^n (n-dimensional Euclidean space).

This can be understood by the following examples:

(i) Consider the linear equation $2x_1 - x_2 = 3$ in two dimensional space R^2. We know that it represents a st. line in R^2.

The points (1, 2), (2, 4), (3, 6), (4, 8), ... satisfy $2x_1 - x_2 = 0$. So this line may be considered as a set of those points (x_1, x_2) which satisfy $2x_1 - x_2 = 3$. This set of points can be written as

$$S = \{(x_1, x_2) \mid 2x_1 - x_2 = 3\}.$$

(ii) Consider the circle

$$x_1^2 + x_2^2 = 36 \text{ in } R^2$$

clearly, the points (1, 3), (2, 3), (3, 2), (3, 3) ... lie inside this circle.

Thus the points (x_1, x_2) of this set satisfy $x_1^2 + x_2^2 < 36$.

The set of points can be written as

$$S = \{(x_1, x_2) \mid x_1^2 + x_2^2 < 36\}$$

3.3. NEIGHBOURHOOD

The $\in$-neighbourhood of a point $\vec{a}$ is the sets of all points $x \in R^n$ whose distance from $\vec{a}$ is less than $\in$, where $\in$ is any positive number, however small. In other words,

$$S = \{x : |x - a| < \in\}$$

Examples: (i) $a \in R^2$ (two dimensional space)

Let $$\vec{a} = a_1\vec{i} + a_2\vec{j} = (a_1, a_2)$$
$$\vec{x} = x_1\vec{i} + x_2\vec{j} = (x_1, x_2)$$

Now, $$\left|\vec{x} - \vec{a}\right| = \in$$
$$|(x_1 - \vec{a}_1)\vec{i} + (x_2 - a_2)\vec{j}| = \in$$
or, $$\sqrt{(x_1 - a_1)^2 + (x_2 - a_2)^2} = \in$$
or, $$(x_1 - a_1)^2 + (x_2 - a_2)^2 = \in^2$$

which represents a circle with centre (a_1, a_2) and radius $\in$.

Thus $\left|\vec{x} - \vec{a}\right| = \in$ represents a circle whose centre is a and radius $\in$.a and x are position vector of A and P respectively; $\in (= AB)$

$$\vec{AP} = \vec{OP} - \vec{OA} = \vec{x} - \vec{a}$$

$|x - a|$ is the distance of P from A. Thus $|x - a|$ represents the interior point of the circle of radius $\in$.

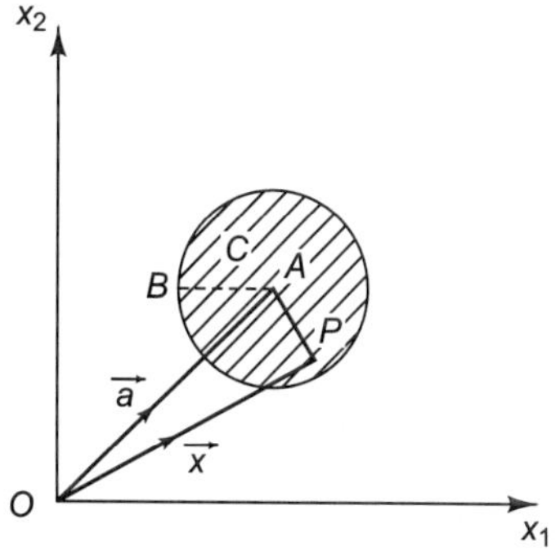

Fig. 3.3

(ii) Similarly, if $a \in R^3$ (three dimensional space), that is,

$\vec{a} = (a_1, a_2, a_3)$, then the neighbourhood of a is the set of an interior points

$\vec{x} = (x_1, x_2, x_3)$ such that
$$\left|\vec{x} - \vec{a}\right| < \in$$

i.e., the distance of x from a is less than $\in$.

i.e. $\sqrt{(x_1 - a_1)^2 + (x_2 - a_2)^2 + \ldots(x_n - a_n)^2} < \in$.

Thus $\left|\vec{x} - \vec{a}\right| < \in$ represents the interior of space of radius $\in$ on three dominational space R^3.

3.4. INTERIOR POINT

A point a is an interior point of set S if there exists an $\in$-neighbourhood about a which contains only points of the set S.

Note: A interior point of S must be an element of S.

3.5. BOUNDARY POINT

A point $\vec{a}$ is boundary point of the set S if every $\in$-neighbourhood about $\vec{a}$, $\in > 0$ (however small) contains points which are in the set and points which are no. in the set.

Note: A boundary point of S must be an element of S.

3.6. OPEN SET

A set S is said to be an open set if it contains only interior points.

3.7. CLOSED SET

A set S is said to be closed set if it contains all its boundary points.

3.8. LINEAR COMBINATION OF VECTORS

Given two vectors $\vec{a}$ and $\vec{b}$ we can always perform operations of addition and scalar multiplication to obtain vectors of the form $\vec{a}+\vec{b}$, $\vec{a}-\vec{b}$, $3\vec{a}+2\vec{b}$, $4\vec{a}-6\vec{b}$ etc. Such vectors are called linear combination of $\vec{a}$ and $\vec{b}$. Thus if $(\vec{a_1},\vec{a_2},\vec{a_3},\ldots\vec{a_r})$ is a set of r vectors from R^n and $\{\lambda_1, \lambda_2, \ldots, \lambda_r\}$ is a set of r scalars, then the vector.

$$\lambda_1 \vec{a_1} + \lambda_2 \vec{a_2} + \ldots + \lambda_r \vec{a_r}$$

is called linear combination of the given set of vectors $\vec{a_1},\vec{a_2},\ldots\vec{a_r}$.

3.9. LINEAR DEPENDENCE AND INDEPENDENCE OF VECTORS

A set of vector $\vec{a_1},\vec{a_1},\ldots\vec{a_r}$ of R^n is said to be Linearly dependent if there exists scalars $\lambda_1, \lambda_2, \ldots \lambda_r$ not all zero, such that,

$$\lambda_1 \vec{a_1} + \lambda_2 \vec{a_2} + \ldots + \lambda_r \vec{a_r} = 0$$

When it is not possible to have the above equality without all λ_i $(= \lambda_1, \lambda_2 \ldots \lambda_r)$ being zero, then the given set of vectors $\vec{a_1},\vec{a_2},\vec{a_r}$ is said to be Linearly Independent.

The following examples make the point clear.

Example 1.

Consider, the vectors $\vec{a_1} = i + 2j = (1, 2)$

$$\vec{a_2} = 2i + 4j = (2, 4)$$

where, i and j are unit vectors

Take, $\lambda_1 = 2, \lambda_2 = -1$, then

$$\begin{aligned}\lambda_1 \vec{a_1} + \lambda_2 \vec{a_2} &= 2(i + 2j) + (-1)(2i + 4j)\\ &= 2i + 4j - 2i - 4j\\ &= 0\end{aligned}$$

Thus the exist $\lambda_1 = 2$ and $\lambda_2 = -1$ for which $\lambda_1 \vec{a}_1 + \lambda_2 \vec{a}_2 = 0$. Therefore, $\vec{a}_1$ and $\vec{a}_2$ are linearly dependent.

Example 2.

The vectors

$$\vec{e}_1 = (1, 0, 0)$$
$$\vec{e}_2 = (0, 1, 0)$$
$$\vec{e}_3 = (0, 0, 1)$$

are linearly independent, since $\lambda_1 \vec{e}_1 + \lambda_2 \vec{e}_2 + \lambda_3 \vec{e}_3 = \vec{0}$

$$\Rightarrow \quad \lambda_1 (1, 0, 0) + \lambda_2 (0, 1, 0) + \lambda_3 (0, 0, 1) = (0, 0, 0)$$
$$\Rightarrow \quad (\lambda_1, \lambda_2, \lambda_3) = (0, 0, 0)$$
$$\Rightarrow \quad \lambda_1 = 0, \lambda_2 = 0, \lambda_3 = 0$$

Note: (i) If a set of vectors is linearly independent, then any subset of it is also linearly independent.

(ii) If a set of vectors is linearly dependent, then a super set of it is also dependent.

3.10. BASIC SET

A linearly independent set of vectors $\vec{a}_1, \vec{a}_2, \ldots \vec{a}_r$ of R^n, having the property, that any vector of R^n can be expressed as a linear combination of these vectors, is called a basic set for R^n.

3.11. LINE-SEGMENT

Geometrically, a portin PQ of the line AB is called line-segment.

Take $P(x_1, y_1)$ and $Q(x_2, y_2)$ two points in R^2 (i.e. two dimensional space.) Equation of any line through these two points are

$$\frac{y - y_2}{y_1 - y_2} = \frac{x - x_2}{x_1 - x_2} = \lambda$$

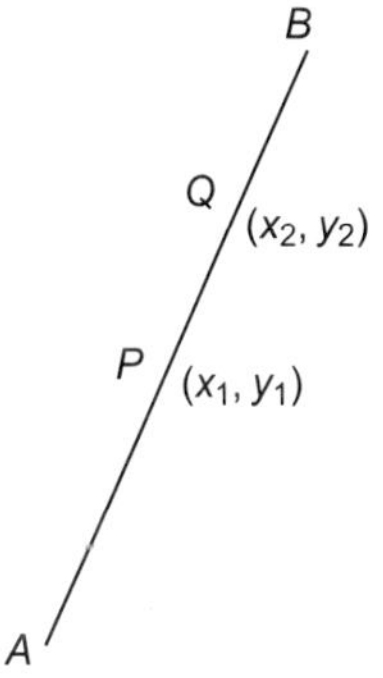

Fig. 3.4

From this, $\dfrac{x - x_2}{x_1 - x_2} = \lambda$

or, $\quad x - x_2 = \lambda (x_1 - x_2)$

or, $\quad x = \lambda x_1 + x_2 - \lambda x_2 \quad$ ...(i)

or, $\quad x = \lambda x_1 + (1 - \lambda) x_2 \quad$ ...(ii)

Similarly $\quad y = \lambda y_1 + (1 - \lambda) y_2$

Let us introduce the three vectors

$$\left.\begin{aligned} v &= (x, y) \\ v_1 &= (x_1, y_1) \\ v_2 &= (x_2, y_2) \end{aligned}\right\} \qquad \text{...(iii)}$$

With the help of (iii), (i) and (ii) can be combined into single equation

$$v = \lambda v_1 + (1 - \lambda)v_2 \qquad \text{...(iv)}$$

Putting $\lambda = 1$ in 4, we get

$$v = v_1 = (x_1, y_1) \text{ i.e. } P$$

Putting $\lambda = 0$ in (iv), we get

$$v = v_2 = (x_2, y_2) \text{ i.e. } Q$$

So for the line-segment PQ; we must have $0 \le \lambda \le 1$

As before, by taking two points $P(x_1, y_1, z_1)$ and $Q(x_2, y_2, z_3)$ in R^3 and using the vectors

$$\vec{v} = (x, y, z)$$
$$\overrightarrow{v_1} = (x_1, y_1, z_1);$$
$$\overrightarrow{v_2} = (x_2, y_2, z_2)$$

we can show that

$$\vec{v} = \lambda \overrightarrow{v_1} + (1 - \lambda) \overrightarrow{v_2} \text{ when } 0 \le \lambda \le 1$$

Extending the above idea for R^n (n-dimentional space), if can be shown that the equation of a line joining two points P and Q in R^n takes the form $v = \lambda v_1 + (1 - \lambda)\, v_2$, where v, v_1, v_2 are each $n \times 1$ column vectors for the line segemnt PQ, must have $0 \le \lambda \le 1$.

The line-segment of the vectors a and b in R^n is the set of vectors.

$$\vec{x} = \lambda \vec{a} + (1 - \lambda)\vec{b}, \text{ where } 0 \le \lambda \le 1$$
$$\vec{x} = (x_1, x_2, \dots x_n)$$
$$\vec{a} = (a_1, a_2, \dots a_n)$$
$$\vec{b} = (b_1, b_2, \dots, b_n)$$

Theorem: Any point on the line-segment joining two points in R^n can be expressed as convex combination of the two points and conversely.

Denote the two points A and B by $\vec{a}$ and $\vec{b}$ respectively. Let C lie on the line segment joining $\vec{a}$ and $\vec{b}$. In the figure the points have been represented by directed line segments (or vectors originating from O).

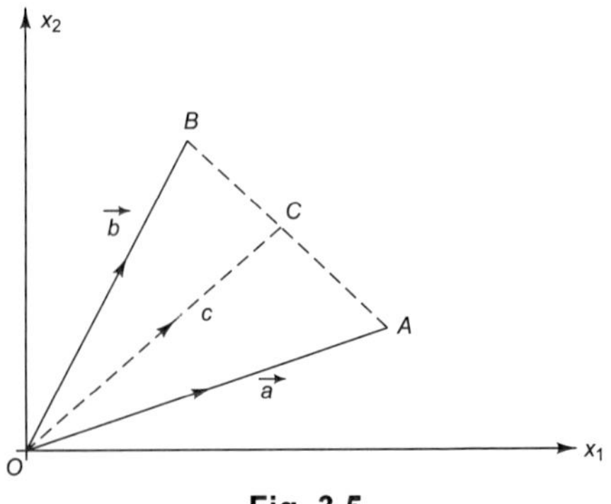

Fig. 3.5

The line segemtn AB is parallel to the line defined by the vector $\vec{b} - \vec{a}$. Now, for some $0 \le \lambda \le 1$, we have

$$\overrightarrow{OA} + \overrightarrow{OC} = \overrightarrow{OC}$$

which means $\vec{a} + \lambda(\vec{b} - \vec{a}) = c$

$$\Rightarrow \quad (1 - \lambda)\,\vec{a} + \lambda\vec{b} = c$$

$$\Rightarrow \quad \vec{c} = \overrightarrow{Ka} + (1 - K)\,\vec{b}$$

where $1 - \lambda = K$, a scalar.

This proves that c is *a* convex.

Combination of $\vec{a}$ and $\vec{b}$.

Converse:

Any point that can be expressed as a convex combination of two points in R^n is on the line segment joining the two points.

Here, given that

$$\vec{c} = K\,\vec{a} + (1 - K)\,\vec{b}$$

or, $\vec{c} - \vec{b} = K\,(\vec{a} - \vec{b})$

where $0 \le K \le 1$

This means that $\vec{c} - \vec{b}$ is some positive multiple of vector $\vec{a} - \vec{b}$. But from the figure,

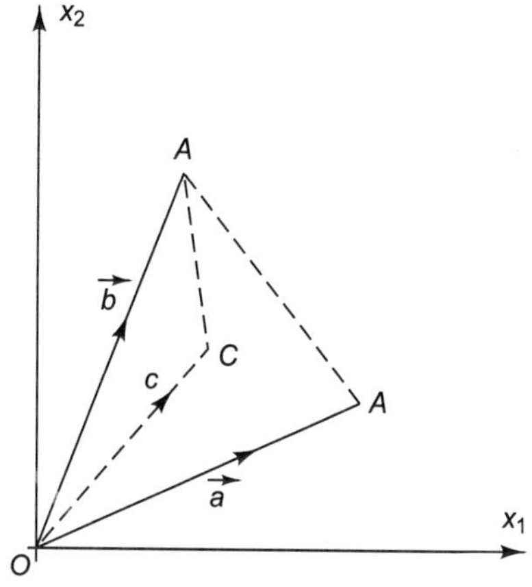

Fig. 3.6

$$\vec{c} - \vec{b} = \overrightarrow{BC}$$

and $\vec{a} - \vec{b} = \overrightarrow{BA}$.

Thus, we find that $\overrightarrow{BC}$ is some positive multiple of $\overrightarrow{BA}$.

That C must lie on the line segment AB. It means C must lie on the line joining $\vec{a}$ and $\vec{b}$.

Note: From above theorem we see that geometrically a convex set is one which contains all the line segments joining two points of the set.

3.12. HYPER PLANE IN R^n (R.U.B.SC., 87)

We know that first degree equation in x, y and z represents a plane in three dimentional space R^3. For examples, $ax + by + cz = d$ represents a plane. If $d = 0$, the plane passes through the origin. In the same way $c_1x_1 + c_2x_2 + c_3x_3 = 0$ where c_1, c_2, c_3 and z all are constants represents place in R^3 as the set of points $x = (x_1, x_2, x_3)$ satisfies $c_1x_1 + c_2x_2 + c_3x_3 = z$.

From the above idea we define the Hyper plane as below

In R^n the set of points $\vec{x} = (x_1, x_2, x_3, \ldots x_n)$ satisfying the equation

$$c_1x_1 + c_2x_2 + c_3x_3 + \ldots + c_nx_n = z, \qquad \ldots(i)$$

not all $c_3 = 0$ is called a hyper plane for given value of $c_1, c_2, \ldots, c_n$ and z.

The equation (i) can be put in the matrix form as

$CX = Z$ where C is the row-vector

$C = [c_1, c_2, c_3, \ldots, c_n]$

X is the Column vector $\begin{bmatrix} x_1 \\ x_2 \\ \ldots \\ x_n \end{bmatrix}$

and $Z = [z]$

Note: (i) If the hyper plane passes through the origin, then its equation is $CX = 0$

(ii) A hyper plane divides R^n into two half spaces which are denoted by

$$X_1 = \{X \mid CX \geq Z\}$$
$$X_2 = \{Z \mid CX \leq Z\}$$

X_1 is the half space i.e. that portion of R^n that containsthe Vector X for which $CX \geq Z$.

X_2 is the half space that contains the vectors X for which$CX \leq Z$.

(iii) But a hyper plane divides the space R^n into three mutually disjoint sets given by

$$X_1 = \{X \mid CX > Z\}$$
$$X_2 = \{X \mid CX = Z\}$$
$$X_3 = \{X \mid CX < Z\}$$

(iv) **Open half spaces**

The sets $X_1 = \{X \mid CX > Z\}$
$X_2 = \{X \mid CX < Z\}$ are called open half spaces

(v) **Closed half spaces**

The sets $X_4 = \{X \mid CX \leq Z\}$
$X_5 = \{X \mid CX \leq Z\}$ are called *closed half spaces*

It is obvious that the hyper planes are closed sets.

We can have the idea of half spaces by the following examples:

Consider two dimensional (R^2) Space

Take $C = (4, -2); Z = 1$ and

$X = (x_1, x_2)$

we have $CX = (4, -2)(x_1, x_2)$

or, $CX = 4x_1 - 2x_2$

Then the two half spaces are given by

$$X_1 = CX \geq Z$$

i.e., $X_1 = 4x_1 - x_2 \geq 1$

and $X_2 = CX \leq Z$

$$X_2 = 4x_1 - 2x_2 \leq 1$$

Putting the Co-ordinates of origin i.e. $x_1 = 0, x_2 = 0$ in $4x_1 - 2x_2 = 1$. We see that X_2 represents the negative side and X_1 positive side of the plane w.r.t. the given line. The points on the line $4x_2 - 2x_2 = 1$ lie on both the half spaces.

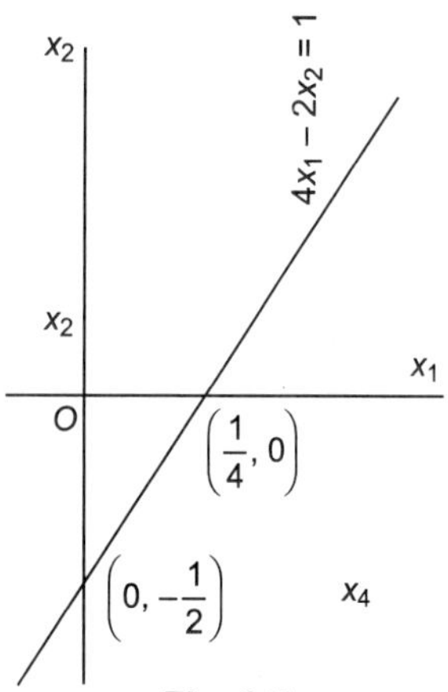

Fig. 3.7

3.13. CONVEX LINEAR COMBINATION

If a and b be two vectors in R^n, then $Z = \lambda a + (1 - \lambda) b, 0 \leq \lambda \leq 1$ is called the convex linear combination of $\vec{a}$ and $\vec{b}$ in R^n and for all convex linear combinations that is for all values of λ between 0 and 1, Z denotes the line segment between $\vec{a}$ and $\vec{b}$ in R^n.

3.14. CONVEX SET

A set S in n-dimensional space R^n is said to be convex if for two points $\vec{a}$ and $\vec{b}$ in S, the line segment joining the two points is also in the set S.

In other words, if $\vec{a} \in S$ and $\vec{b} \in S$ where, $S \subset R^n$ and also if $Z(\lambda) \in S$ for all values of λ when $0 \leq \lambda \leq 1$, $\vec{Z}(\lambda) = \lambda \vec{a} + (1 - \lambda) \vec{b}$, then S is called convex.

Thus a convex subset has no holes, in it. Its boundaries are bent away from the set. The following example from R^2 will help us to convey the idea clearly.

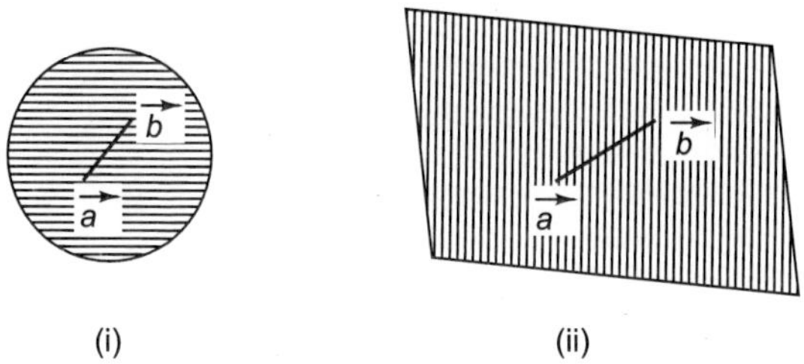

Fig. 3.8 Convex Set

(i) It is the interior and boundary of circle and since the line segment of any two points $\vec{a}$ and $\vec{b}$ is lying inside the circle the space is convex. Clearly, the space has infinite number of boundary points and all of its boundary points are extreme points. Figure (ii) is also convex.

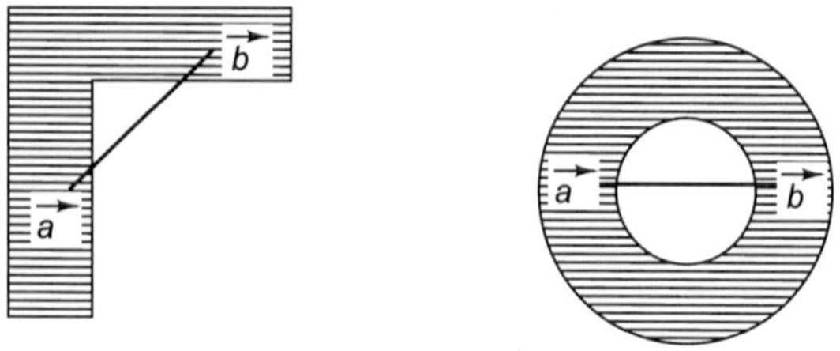

Fig. 3.9 Non-Convex Set

3.15. EXTREME POINT OF A CONVEX SET

[Meerut 70, 76; Delhi 68, 71]

An extreme point (or vector) of convex set is a point of the set which does not lie on any segment joining two other points of the set.

Thus a point x of a convex set S is an extreme point of the set, if there does not exist any pair of points x_1, x_2 in S such that

$$x = \lambda x_1 + (1 - \lambda)x_2,\ 0 < \lambda < 1$$

Note: (i) A boundary point $\vec{a}$ is called an extreme point if there does not exist any segment is S whose centre is the point $\vec{a}$.

(ii) An extreme point is a boundary point of the set; however, not all boundary points of convex set are necessarily extreme points.

Here, the figure is convex. A, B, C, D, E, F, M, all are boundary points. But M is not extreme point.

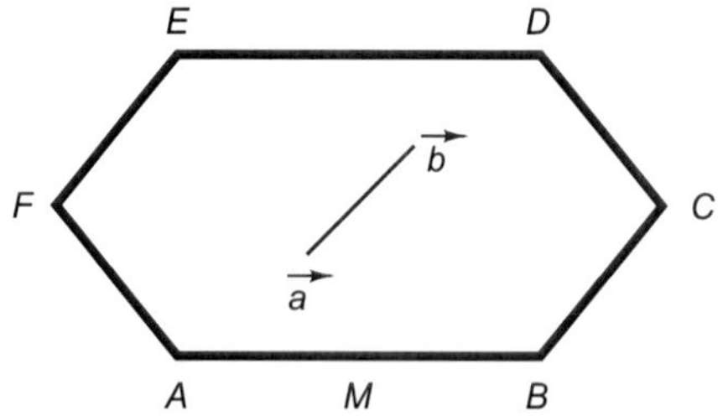

Fig. 3.10

(iii) If a set consists of a single point only then this point is itself an extreme point.

(iv) The adjoining figure is convex, being the space inside a wedge.

If it is a finite (that is OPQ forms a triangle) then O, P and Q are extreme points. If however, the wedge is infinite, then O is the only extreme point (or vertex point).

3.16. UNBOUNDED CONVEX SET

A convex set S is said to be unbounded if for every points $a \in S$, a vector $\vec{b} \neq 0$ can be formed, such that for all values of $\lambda > O_1$ the point $\vec{a} + \lambda\vec{b} \in S$.

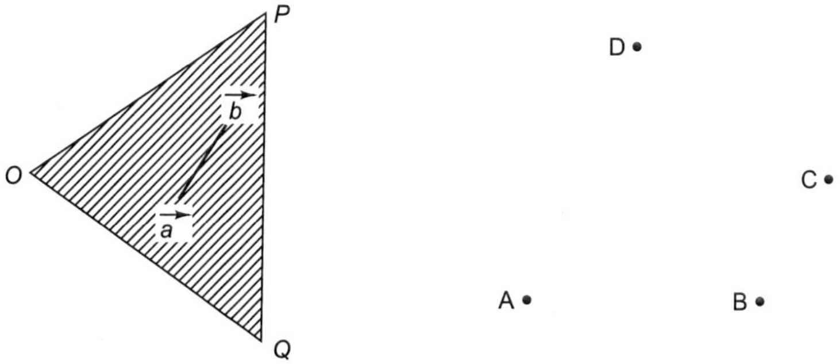

Fig. 3.11

Otherwise, S is called bounded.

From the above definition it follows that S is bounded if there exists $\vec{a}$ point a in S such that for every $\vec{b} \neq 0$ there is a λ such that $\vec{a} + \lambda\vec{b}$ is outside S.

Examples

The figure is the set of discrete points, A, B, C, D; and this collection of set of discrete points is not convex.

3.17. CONVEX COMBINATION

A linear convex combination of a finite number of points $a_1, a_2, ..., a_r$ is defined as

$$\vec{a} = \lambda_1 \vec{a}_1 + \lambda_2 \vec{a}_2 + ... + \lambda_r \vec{a}_r$$

where, $\lambda_1, \lambda_2 ... \lambda_r \geq 0$, and

$$\lambda_1 + \lambda_2 + ... + \lambda_r = 1$$

3.18. CONVEX HULL

Let X be an arbitrary set in R^n.

Then the convex hull $C(X)$ of any given set of points X is the set of all convex combinations of sets of points from X.

Note: (i) $C(X)$ is the smallest convex set containing set of point X.

(ii) Since X is arbitrary, it may be the union of several sets in R^n. This idea can be well explained by the following figures.

(a) If X_1 and X_2 be the subset in R^2 as shown in the figure, the Convex Hull is given by the shaded region in the figure.

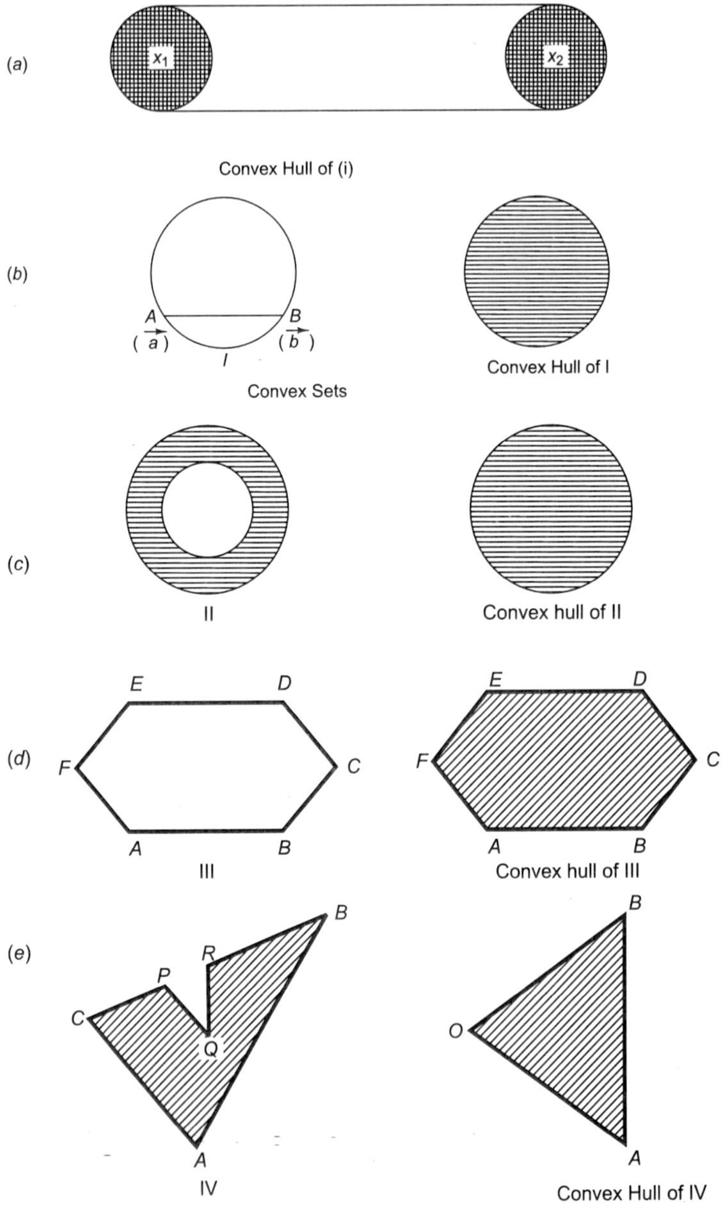

Fig. 3.12

Illustrative Examples

Example 1. Prove that set X defined as

$$Z = \{(x_1, x_2) \mid x_1, x_2 \le 1, x_1 \ge 0, x_2 \ge 0\}$$

is not Convex. Also, defined the convex hull of the set X.

(R.U., B.Sc. '88)

(V.B.U. Part III Hons. 97 (old spl.))

From knowledge of co-ordinate geometry $x_1, x_2 = 1$ represents hyperbola. Here $x_1 \ge 0$, $x_2 \ge 0$. So $x_1 \cdot x_2 = 1$ represents hyperbola in the first quadrant.

In the adjoining figure we denote the set of points defined by the shaded portion. Clearly the set X is not convex. This can further be shown in the following way:

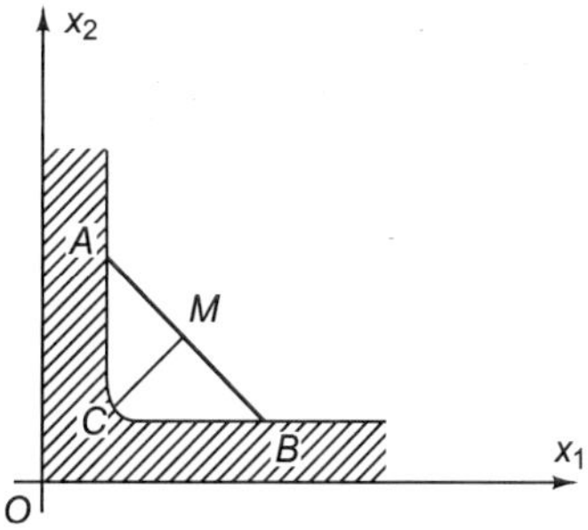

Fig. 3.13

Let $A\left(\alpha, \dfrac{1}{\alpha}\right)$ and $B\left(\beta, \dfrac{1}{\beta}\right)$, $\alpha, \beta > 0$ be two distinct points in the set and if x_1 and x_2 denote A and B then

$$x_1 = \left(\alpha, \frac{1}{\alpha}\right) \quad \text{...(i)}$$

and $$x_2 = \left(\beta, \frac{1}{\beta}\right)$$

The convex combination of x_1 and x_2, (that is the line segment joining x_1 and x_2) is given by the set of points.

$$x = \lambda x_1 + (1 - \lambda)x_2, \; 0 \le \lambda \le 1 \quad \text{...(ii)}$$

If $x = z(l_1, l_2)$, then

$$l_1 = \lambda a + s\,(1 - \lambda)\,\beta$$

and $$l_2 = \lambda \cdot \frac{1}{\alpha} + (1 - \lambda) \cdot \frac{1}{\beta}$$

If $\lambda = \dfrac{1}{2}$, then middle point M of AB is given by,

$$l_1 = \frac{1}{2}\alpha + \left(1 - \frac{1}{2}\right)\beta = \frac{\alpha + \beta}{2}$$

and $$l_2 = \frac{1}{2} \cdot \frac{1}{\alpha} + \left(1 - \frac{1}{2}\right) \cdot \frac{1}{\beta} = \frac{1}{2}\left(\frac{1}{\alpha} + \frac{1}{\beta}\right)$$

Clearly, $l_2 > \dfrac{1}{l_1}$ for $l_2 > \dfrac{1}{l_1}$

$$= > \quad \frac{1}{2}\left(\frac{1}{\alpha}+\frac{1}{\beta}\right) > \frac{1}{\frac{\alpha+\beta}{2}}$$

$$= > \quad \frac{1}{2}\,\frac{\alpha+\beta}{\alpha\beta} > \frac{2}{\alpha+\beta}$$

$$= > (\alpha+\beta)^2 > 4\alpha\beta$$

$$= > \alpha^2+\beta^2+2\alpha\beta-4\alpha\beta > 0$$

$$= > (\alpha-\beta)^3 > 0 \text{ which is true}$$

Thus $l_2 > \frac{1}{l_1}$ i.e. $l_1\, l_2 > 1$

$\therefore$ $M\,(l_1, l_2)$ falls outside the set.
Hence, the set is not convex.

Second Part: Since by definition the Convex hull should be the smallest convex set containing X. The whole positive quadrant is the convex hull of the set X. This is so because as $A \to \infty$ ($x_1 = 0$ is an asymptote) and $B \to \infty$ ($x_2 = 0$ is an asymptote), the line segment traces entire positive quadrant. Hence, the result.

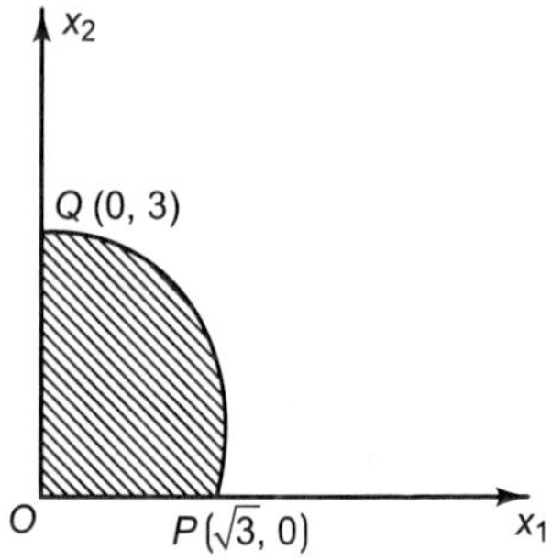

Fig. 3.14

Example 2. Show that the X defined as $X = \{(x_1, x_2) \mid x_2 - 3 \geq -x^2{}_1; x_1 \geq 0, x_2 > 0\}$ is convex. [Meerut 73]

In the figure the points defined by the given set is shown by the shaded portion, clearly the set X is convex.

Example 3. Draw the convex hull of the discrete point set

$$X = \{(0, 0), (7, 0), (4, 4), (0, 5)\}$$

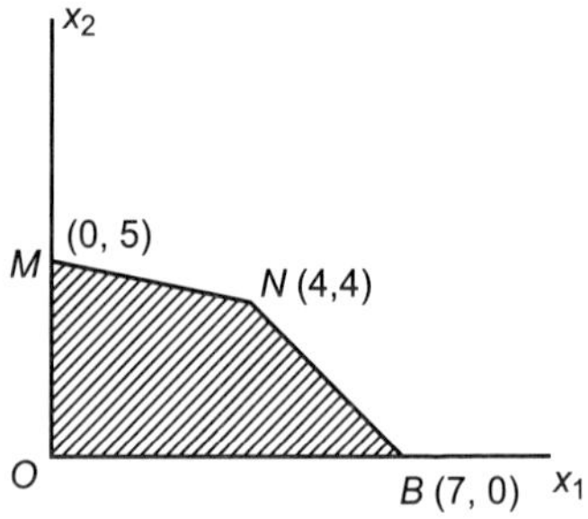

Fig. 3.15

The figure denotes the convex hull of the given set. Since $N\,(4, 4)$ falls on the other side of the line BM relative to O, we get the convex hull by simply joining OB, BN, NM and MO: and indicated by the shaded portion.

Example 4. Draw the convex hull of the discrete point set

$$X = \{(0, 0), (7, 0), (2, 2), (0, 5)\}$$

If we joint OP, PQ, QR and RO, we get the quadrilateral $OPQR$.

Clearly it is not convex and hence not the convex hull of the given point set, however, this non-convex set would be convex if we join PR (that is if we include ΔPQB) and the set of tall point inside the ΔOPR (as shown shaded portion) would be the required convex-hull. We can therefore, write in the case

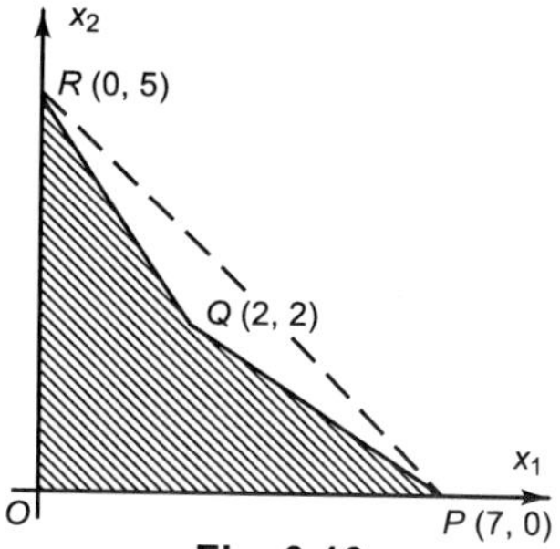

Fig. 3.16

$$C(X) = \{(x_1, x_2) \mid 5x_1 + 7x_2 \leq 35, x_1 \geq 0\},$$

since the equation of the line PR is $5x_1 + 7x_2 = 35$, Obviously $Q(2, 2)$ becomes an interior point of $C(X)$.

3.19. CONVEX POLYHEDRON

A Convex is called a convex polyhedron if it be (i) bounded and (ii) if the number of extreme points be finite.

Alternatively, the set of all convex combinations of a finite number of points is called the *convex polyhedron* spanned by these points.

Note—For convex polyhedron the extreme points are called corner points.

Examples (i) Convex hull of the eight vertices of a cube is a convex polyhedron.

(ii) The figure is a convex polyhedron and A, B, C, D, E, F are its corner points (or vertices or vertices points).

(iii) Clearly, this circle is not a convex polyhedron because it has infinite number of extreme points.

(iv) The set of the area of a triangle is convex polyhedron of its vertices.

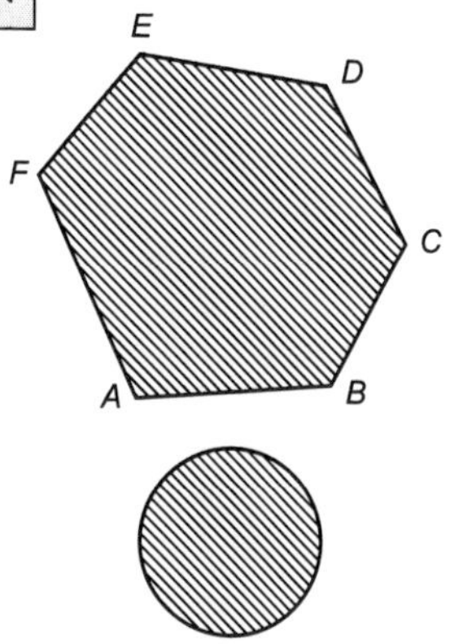

Fig. 3.17

3.20. CONVEX CONE

A Convex cone is one which is convex.

Simplex: A Simplex is an n-dimensional convex polyhedron having exactly $(n + 1)$ vertices.

Examples: (i) A simplex in zero-dimension is a point.

(ii) A simplex in one-dimension is a line.

(iii) A simplex is in two-dimension is a triangle.

(iv) A simplex is three-dimension is a tetrahedron.

SOME IMPORTANT THEOREMS

3.21. SHOW THAT THE SPHERE IS A CONVEX SET

[R.U., B.Sc., 88, V.B.U., Part II Hons. 96 H, Part II Hons. 99 H]

Solution: Let the equation of sphere be

$$x^2 + y^2 + z^2 = r^2 \quad \text{...(i)}$$

Let $\vec{a}$ and $\vec{b}$ be represent any two points $A(x_1, y_1, z_1)$ and $B(x_2, y_2, z_2)$ respectively in the sphere (i).

So, $\vec{a} = (x_1, y_1, z_1)$ and

$\vec{b} = (x_2, y_2, z_2)$

$$\therefore \quad x_1^2 + y_1^2 + z_1^2 \le r^2 \quad \text{...(ii)}$$

and

$$x_2^2 + y_2^2 + z_2^2 \le r^2 \quad \text{...(iii)}$$

Take the convex combination of $\vec{a}$ and $\vec{b}$ be given by

$$\vec{C} = \lambda\vec{a} + (1 - \lambda)\,\vec{b}, \quad 0 \le \lambda \le 1 \quad \text{...(iv)}$$

If $\vec{C} = (x_3, y_3, z_3)$, then

$$\left.\begin{aligned} x_3 &= \lambda x_1 + (1 - \lambda)\,x_2 \\ y_3 &= \lambda y_1 + (1 - \lambda)\,y_2 \\ z_3 &= \lambda z_1 + (1 - \lambda)\,z_2 \end{aligned}\right\} \quad \text{...(v)}$$

Obviously (v) gives the co-ordinates of any point on the line joining the points $A\,(x_1, y_1, z_1)$ and $B\,(x_2, y_2, z_2)$.

Now, $x_3^2 + y_3^2 + z_3^2 = \{\lambda x_1 + (1 - \lambda)\,x_2\}^2 + \{\lambda y_1 + (1 - \lambda y_2\}^2 + (\lambda z_1 + (1 - \lambda)z_2\}^2$

or, $x_3^2 + y_3^2 + z_3^2 = \lambda^2\,(x_1^2 + y_1^2 + z_1^2) + (1 - \lambda)^2\,(x_2^2 + y_2^2 + z_2^2) + 2\lambda\,(1 - \lambda)\,(x_1\,x_2 + y_2\,y_2 + z_1\,z_2)$

$$< \lambda^2 + (1 - \lambda)r^2 + 2\lambda\,(1 - \lambda)\,\sqrt{x_1^2 + y_1^2 + z_1^2}\,\sqrt{x_2^2 + y_2^2 + z_2^2}$$

by (ii) and (iii) and using Cauchy-inequality.

$$= \lambda^2\,r^2 + (1 - \lambda)^2\,r^2 + 2\lambda\,(1 - \lambda)\,r^2$$

$$= r^2\,(\lambda^2 + (1 - \lambda^2) + (2\lambda - 2\lambda^2)$$

$$= r^2\,(\lambda^2 + 1 + \lambda^2 - 2\lambda + 2\lambda - 2\lambda^2) = r^2$$

Thus $\quad x_3^2 + y_3^2 + z_3^2 < r^2$.

So $C = (x_3, y_3, z_3)$ is in the sphere (i). But this point is on the line-segment represented by the line joining the points

$$A\,(x_1, y_1, z_1) \text{ and } B\,(x_2, y_2, z_2).$$

But this is an arbitrary point.
Hence, the line-segment represented by (iv) belongs the set.
Hence the sphere is a convex set.

3.22. Theorem I: *A hyper plane is convex set.*

Proof: Let the hyper plane be

$$X = \{x: Cx = z\}$$

Let, x_1 and x_2 be any line points in the hyperplane X.

$$\therefore \quad Cx_1 = z \quad \text{...(i)}$$

$$\text{and} \quad Cx_2 = z \quad \text{...(ii)}$$

If $\quad x_3 = \lambda x_1 + (1 - \lambda)\, x_2,\ 0 \le \lambda \le 1$, then

$Cx_3 = \lambda Cx_1 + (1 - \lambda)\, Cx_2$

$= \lambda z + (1 - \lambda)\, z$, by (i) and (ii)

$Cx_3 = z$, which implies that,

$x_3 = \lambda x_1 + (1 - \lambda)\, x_2$ is also a point is in the hyper plane X.
Therefore, by definition, the hyper plane X is a convex set.

3.23. Theorem II: *Intersection of two convex sets is also a convex set*

(R.U.B.Sc., 85, 90 H, V.B.U. Part III Hons. 98 H)

Proof: Let, X_1 and X_2 be two convex sets.
Let, the set X_3 be the intersection of X_1 and X_2

i.e. $\quad X_3 = X_1 \cap X_2$.

If $\quad x_1 \in X_1 \cap X_2$, then $x_1 \in X_1$ and $x_1 \in X_2$

If $\quad x_2 \in X_1 \cap X_2$, then $x_2 \in X_1$ and $x_2 \in X_2$

Since X_1 and X_2 are convex sets; therefore;

$$x_1, x_2 \in X_1 \Rightarrow \lambda x_1 + (1 - \lambda)\, x_2 \in X_1,\ 0 \le \lambda \le 1$$

$$\text{and} \quad x_1, x_2 \in X_2 \Rightarrow \lambda x_1 + (1 - \lambda)\, x_2 \in X_2,\ 0 \le \lambda \le 1$$

Since, $\quad \lambda x_1 + (1 - \lambda)\, x_2 \in X_1$ and

$\lambda x_1 + (1 - \lambda)\, x_2 \in X_2$,

therefore, $\lambda_1 + (1 - \lambda)\, x_2 \subset X_1 \cap X_2,\ 0 \le \lambda \le 1$

i.e. $\quad \lambda x_1 + (1 - \lambda)\, x_2 \in X_3$

Hence, by definition,

$$X_3 = X_1 \cap X_2 \text{ is a convex set.}$$

3.24. Theorem III: *The set of all feasible solution of a L.P.P. is a convex set.*

[Meerut 68, 69, O.R. 74; Delhi 68, 69, 71, S.I. 71, R.U.M.Sc. 83]

Proof: Let X be the set of all feasible solutions of a L.P.P;

$$Ax = b,\ x \ge 0$$

Case I: When X has only one element.

If the set X has only one element, then X is convex set. The theorem is true in this case.

Case II: When X has at least two elements.
Let x_1 and x_2 be any two distinct elements is X.

$$\therefore \qquad Ax_1 = b,\ x_1 \geq 0 \qquad \text{...(ii)}$$

$$\text{and} \qquad Ax_2 = b,\ x_2 \geq 0 \qquad \text{...(iii)}$$

$$\text{If} \qquad x_3 = \lambda x_1 + (1-\lambda)x_2,\ 0 \leq \lambda \leq 1,$$

$$\text{then} \qquad Ax_3 = \lambda A x_1 + (1-\lambda)\, Ax_2$$

$$= \lambda b + (1-\lambda)\, b, \text{ from (ii) and (iii)}$$

$$= b$$

Since, $0 \leq \lambda \leq 1,\ 1-\lambda \leq 0$
Now, $x_1 \geq 0,\ x_2 \geq 0,\ \lambda \geq 0,\ 1-\lambda \geq 0,$

$$\therefore \qquad x_3 = \lambda x_1 + (1-\lambda)x_3 \geq 0$$

This shows that x_2 satisfies (1).

Thus $x_3 = \lambda x_1 + (1-\lambda)\, x_2$ is also feasible solution and so belongs to X. But x_3 is convex combination of any two distinct point x_1 and x_2 in X. Hence by definition the set X is convex set.

3.25. Theorem IV: *The set of all convex combinations of a finite number of point $x_1, x_2, \ldots x_n$ is a convex set.*

[Meerut73, S.R.U.M.Sc. 81, V.B.U. Part III Hons. 98]

Proof: Let X be a set of all convex combinations of a finite number of points,

$$\text{i.e.} \quad X = \left\{ x : x = \sum_{j=1}^{n} \lambda_j\, x_j,\ \sum_{j=1}^{n} \lambda_j = 1, \lambda_j \geq 0 \right\}$$

Let $u, v \in X$. Then

$$u = \sum_{j=1}^{n} a_j x_j,\ \sum_{j=1}^{n} a_j = 1, \quad a_j \geq 0$$

$$\text{and} \quad v = \sum_{j=1}^{n} b_j x_j,\ \sum_{j=1}^{n} b_j = 1, \quad b_j \geq 0$$

Let us consider

$$W = \lambda u + (1-\lambda)\, v,\ 0 \leq \lambda \leq 1$$

$$\therefore \qquad W = \lambda \sum_{j=1}^{n} a_j\, x_j + (1-\lambda) \sum_{j=1}^{n} b_j\, x_j$$

$$= \sum_{j=1}^{n} \{\lambda\, a_j + (1-\lambda)\, b_j\}\, x_j$$

$$= \sum_{j=1}^{n} c_j\, x_j, \text{ where } c_j = \lambda a_j + (1-\lambda)\, b_j$$

$$\text{Now,} \quad \sum_{j=1}^{n} c_j = \sum_{j=1}^{n} [\lambda a_j + (1-\lambda) b_j]$$

$$= \lambda \sum_{j=1}^{n} a_j + (1-\lambda) \sum_{j=1}^{n} b_j$$

$$= \lambda \cdot 1 + (1-\lambda) \cdot 1 = 1.$$

Also, $c_j = \lambda a_j + (1-\lambda) b_j \geq 0, \forall j$

Hence, $W = \lambda u + (1-\lambda) v \in X. \ 0 \leq \lambda \leq 1.$

Hence, by definition X is a convex set.

3.26. Theorem V: *The closed half spaces*

$$H_1 = \{x = 1 \ cx \geq z\}$$

and $H_2 = \{x \mid cx \geq z\}$ are convex sets.

Proof: Let x_1, x_2 be any two points of H_1. Then

$$cx_1 \geq z \text{ and } cx_2 \geq z$$

If $0 \leq \lambda \leq 1$, then $c[\lambda x_1 + (1-\lambda) x_2]$

$= \lambda cx_1 + (1-\lambda) cx_2 \geq \lambda z + (1-\lambda) z$

i.e. $C[\lambda x_1 + (1-\lambda) x_2] \geq z$

But $x_1, x_2 \in H_1, 0 \leq \lambda \leq 1$

$\therefore \lambda x_1 + (1-\lambda) x_2 \in H_1$

Hence, H_1 is a convex set. If $x_1, x_2 \in H_2, 0 \leq \lambda \leq 1$, then replacing $\geq$ by $\leq$ in the above results, we get

$$\lambda x_1 + (1-\lambda) x_2 \in H_2$$

Hence, H_2 is also convex set.

Theorem: A hyper plane is a closed set.

Proof: Let $cx = H$ be a hyper plane. We shall show that there can be no boundary point of the hyper plane which does not belong to the hyper plane.

Let Z be a boundary point of the hyper plane and suppose that it is not a point of the hyper plane.

Then either $z > H$ or $cz > H$

Let $cz < H$ and $cz = a < H$

Now $\quad cx = c(z + x - z)$

$$= cz + c(x - z)$$

$$\Rightarrow \quad |cx| \leq |cz| + |c(x-z)|$$

$$\leq a + |c| \ |x - z| \quad \text{...(i)}$$

Consider the *nbd* of z defined by $|x - z| < \in$ where $\in$ is an arbitrary small positive number.

Letting $\quad \in = \dfrac{H - a}{2|c|}$

We get from (i),

$$|cx| < a + |c| \in = a + |c| \frac{H-a}{2|c|} = a + \frac{H-a}{2}$$

$$= \frac{H+a}{2} < H$$

This shows that x is in the half space $cx < H$. Hence there exists a *nbd* of z which contains no point of the hyper plane.

Therefore z is not a boundary point of the hyper plane. Hence we arrive at contradiction. Thus there is no boundary point of the hyper plane which is not in the plane.

Hence the hyper plane is a closed set.

3.27. Theorem VI: *If the convex set of the feasible solution of AX = b, x ≥ 0 is a convex polyhedron, then atleast one of the extreme point gives an optimal solution.*

[Meerut 70; Delhi 68]

Proof: Let $x_1, x_2 \dots x_t$ be the extreme point of the set X of all the feasible solutions of $Ax = b$, $x \geq 0$. Let Z the objective function which is to be maximized by given by $Z = CX$.

If, $x_m \in X$ is the optimal solution, then Max. $Z = cx_m$.

Now, if x_m is an extreme point, then the theorem is proved.

If X_m is not an extreme point in X.

Since X is convex polyhedron, therefore x_m can be expressed as a convex combination of the extreme point of X.

Since X is convex polyhedron, therefore x_m can be expressed as a convex combination of the extreme point of X.

i.e. $$x_m = \lambda_1 x_1 + \lambda_2 x_2 + \dots + \lambda_t x_t,$$

$$\lambda_i \geq 0, \sum_{i=1}^{t} \lambda_j = 1$$

Let Z_m be the maximum value of Z.

$$\therefore \quad Z_m = cx_m = c(\lambda_1 x_1 + \lambda_2 x_2 + \dots + \lambda_t x_t) \quad \dots(i)$$

$$= \lambda_1 cx_2 + \lambda_2 cx_2 + \dots + \lambda_t cx_t$$

If maximum of cx_j is cx_p, then

$$z_m \leq (\lambda_1 + \lambda_2 + \dots + \lambda_t)\, cx_p$$

i.e. $Z_m \leq cx_p$ as $\sum_{j=1}^{t} \lambda_j = 1$

But Z_m is the maximum value of Z. Therefore

$$Z_m = cx_p \quad \text{or} \quad cx_m = cx_p$$

$$\therefore x_m = x_p \text{ (one of the extreme points)}$$

Hence, the optimal solution is attained at the extreme point. Hence, the theorem.

3.28 Theorem VII: *If A is any finite subset of vectors in R^n, then the convex hull of A is the set of convex combinations of vectors in A.*

Proof: Let A be a finite subset of vector in R^n and $<A>$ be its convex hull.

Let S be the set of all convex combinations of vectors in A. Then we know that S is a convex set containing A; and hence $< A > CS$.

Again, $< A >$ contains A. We shall show that this implies that $<A> \supset S$. We use finite induction on k, the number of vectors in A, to prove this result.

For $\kappa = 2$, let x_1, x_2 be two vectors in A. Then $< A >$ contains x_1, x_2 and is convex set.

$$\therefore \quad x = \lambda x_1 + (1 - \lambda)x_2 \cdot 0 \leq \lambda \leq 1$$

is *a* so in $< A >$.

$\therefore$ All possible convex combinations of x_1 and x_2 are in $< A >$.

Assume, that for any positive integer κ, the result is true for a set of atmost $\kappa - 1$ vectors.

Consider now the set

$$A = \{x_1, x_2, \ldots x_n\}$$

Then $< A >$ is convex set containing A, and in particular

$$\{x_1, x_2, \ldots x_{k-1}\}.$$

Let $$X_1 = \left[x \mid x = \sum_{i=1}^{\kappa-1} \mu_i x_i, \mu_i \geq 0, \sum_{i=1}^{\kappa-1} \mu_i = 1 \right]$$

Thus by induction hypothesis

$$< A > \supset X,$$

Now $< A >$ contains X_1 as well as $\{x_\kappa\}$

$\therefore$ $< A >$ contains all segments joining x_κ to a point of

or, $$x = \delta x_\kappa + (1 - \delta) \sum_{i=1}^{\kappa-1} \mu_i x_i, \mu_i \geq 0, \sum_{i=1}^{\kappa-1} \mu_i = 1$$

is a point in $< A >$.

Now,

$$x = \delta x_\kappa + (1 - \delta) \sum_{i=1}^{\kappa-1} \mu_i x_i, \mu_i \geq 0, \sum_{i=1}^{\kappa-1} \mu_i = 1$$

$$\Rightarrow \qquad x = \sum_{i=1}^{\kappa} \beta_i x_i,$$

Where, $\beta_i = (1 - \delta) \mu_i, \beta_\kappa = \delta$ for $i = 1, 2, \ldots, \kappa - 1$

Since $\mu_i \geq 0$ for each i, and $0 \leq \delta \leq 1$,

$\therefore$ $\beta_i \geq 0$ for $j = 1, 2, \ldots, \kappa$

Also $$\sum_{j=1}^{n} \beta_j = \delta + (1 - \delta), \sum_{i=1}^{n} \mu_i = 1$$

$$\therefore \quad x = \sum_{i=1}^{\kappa} \beta_i x_i$$

is a convex combination of the vectors $x_1, x_2, \ldots x_\kappa$ and is in $< A >$.

$$\therefore \quad < A > \supset S.$$

We conclude that $< A > = S$.

In view of above theorem, the convex polyhedron may be equivalently defined as the convex hull of a set of points S, consisting of a finite number of points in R^n. For examples, the convex hull of the eight vertices of a cube is a convex polyhedron.

Problem on Convex Combination

Determine the interior point or boundary point and express them as convex combination of the extreme points of the following set of points

$$(0, 0), (1, 0), (1, 2), (1, 1) \text{ and } (4, 0)$$

On plotting, we see that (1, 1) is the interior point and (1, 0) is the boundary point.

Let $(1, 1) = \lambda_1 (0, 0) + \lambda_2 (1, 2) + \lambda_3 (4, 0)$

where $\lambda_1, \lambda_2, \lambda_3 \geq 0$ and $\lambda_1 + \lambda_2 + \lambda_3 = 1$

or, $(1, 1) = (0, 0) + (\lambda_2, 2\lambda_2) + (4\lambda_3, 0)$

$= (0 + \lambda_2 + 4\lambda_3, 0 + 2\lambda_2 + 0)$

On equating,

$$1 = \lambda_2 + 4\lambda_3 \quad \text{...(i)}$$

and

$$1 = 2\lambda_2 \quad \text{...(ii)}$$

From (ii), $\lambda_2 = \frac{1}{2}$

Putting $\lambda_2 = \frac{1}{2}$ in (i) $\lambda_3 = \frac{1}{8}$

But $\lambda_1 + \lambda_2 + \lambda_3 = 1$

$$\lambda_1 = 1 - \lambda_2 - \lambda_3 = 1 - \frac{1}{2} - \frac{1}{8} = \frac{3}{8}$$

$$\therefore \quad (1, 1) = \frac{3}{8} (0, 0) + \frac{1}{2} (1, 2) + \frac{1}{8} (4, 0)$$

Again, for the boundary point (1, 0), we take

$$(1, 0) = \lambda_1 (0, 0) + \lambda_2 (1, 2) + \lambda_3 (4, 0)$$

$$= (0, 0) + (\lambda_2, 2\lambda_2) + (4\lambda_3, 0)$$

$$= (0 + \lambda_2 + 4\lambda_3, 0 + 2\lambda_2 + 0)$$

$$\therefore \quad 1 = \lambda_2 + 4\lambda_3 \quad \text{...(iii)}$$

and

$$0 = 2\lambda_2 \quad \text{...(iv)}$$

From (iv), $\lambda_2 = 0$

From (iii), $\lambda_3 = \frac{1}{4}$

But $$\lambda_1 + \lambda_2 + \lambda_3 = 1$$

$$\lambda_1 + 0 + \frac{1}{4} = 1$$

$$\lambda_1 = \frac{3}{4}$$

$\therefore$ $$(1, 0) = \frac{3}{4}(0, 0) + \frac{1}{4}(4, 0)$$

EXERCISE

1. Indicate whether the following statements are true or false.
   (a) All boundary points are extreme points.
   (b) All extreme points of a convex set are its boundary points.
   (c) The vertices are boundary points and the converse is true.
2. Which of the following sets are convex?
   (all $x_1, x_2 \in R^2$)
   (a) $X = \{(x_1, x_2) \mid 4x_1^2 + x_2^2 \leq 36\}$
   (b) $X = \{(x_1, x_2) \mid x_1 \leq 4, x_2 \geq 2\}$
   (c) $X = \{(x_1, x_2) \mid 2x_1 + x_2 \leq 19, 2x_1 + 3x_2 \leq 22\}$
   (d) Set of points on the straight line
   $$y = mx + c$$
   (e) Points (0, 0), (0, 1), (1, 0) and (1, 1).
3. Sketch the convex hull generated by the following set of points. Determine the interior or boundary points if any, and express them as a convex combination of the extreme points:
   (a) (0, 0), (1, 0), (1, 1), (0, 1)
   (b) (0, 0), (0, 1), (1, 2), (1, 1), (4, 0) [Delhi '76]
   (c) (0, 0), (1, 0), (1, 2), (1, 1), (4, 0)
4. Determine the convex hull of the following set X.
   (a) $X = \{(x_1, x_2) \mid x_1^2 + x_2^2 = 1\}$
   (b) $X = \{(x_1, x_2)\}$. [Delhi '72, '75]
5. The open half space,
   $\{x \mid cx > z\}$ and $\{x \mid cx < z\}$ are convex set.
6. Prove that the set
   $X = \{(x_1, x_2) \mid x_1^2 + x_2^2 \leq 4\}$ is a convex set.
   [Delhi B.Sc. 1977]
7. Sketch the convex polygon spanned by the following points in a two dimensional Euclidean space (R^2). Which of these points are vertices? Express the other as the convex linear combination of the vertices.

$$(0, 0), (0, 1)\ (1, 0), \left(\frac{1}{2}, \frac{1}{4}\right)$$

8. Consider the triangle with vertices (0, 0), (2, 0), and (1, 1). Express the points (1, 0.5), (0.5, 0.5) as a convex combination of the extreme points.
9. Prove that the following sets are convex.
   (i) The interior of a circle
   (ii) The interior of a rectangle
   (iii) The interior of a triangle
10. Prove that both the sets A and B as defined below,
    $$A = \{(x_1, x_2) \mid 2 \le x_1 \le 3, 2 \le x_2 \le 6\}$$
    $$B = \{(x_1, x_2) \mid 2 \le x_1 \le 6, 2 \le x_2 \le 3\}$$
    are convex but $A \cup B$ is not convex.
11. Show the sets as defined below,
    $$A = \left\{\begin{array}{r}(x_1, x_2) \mid -2x_1 + 6x_2 \le 12, \\ -3x_1 + 5x_2 \ge 6, \\ x_1 \ge 0,\ x_2 \ge 0\end{array}\right\}$$
    $$B = \{(x_1, x_2) \mid -2x_1 + 6x_2 \ge 12, . \\ -3x_1 + 5x_2 \le 6,\ x_1 \ge 0,\ x_2 \ge 0\}$$
    are convex but $A \cup B$ is not convex.

ANSWERS

1. (a) False (b) True (c) False
2. (a) Convex (b) Convex (c) Convex
   (d) Not Convex (e) Convex.
3. (b) (1, 1) is the interior points of the convex hull.

   Also, $[1, 1] = \frac{3}{5}\ [0, 1] = \frac{1}{5}\ [1, 2] + \frac{1}{5}\ [4, 0]$

   (c) (1, 0) is the boundary point and (1, 1) is the interior point.

   Also, $[1, 0] = \frac{3}{4}\ [0, 0] + \frac{1}{4}\ [4, 0]$

   $[1, 1] = \frac{3}{4}\ [0, 0] + \frac{1}{2}\ [1, 2] + \frac{1}{8}\ [4, 0]$
4. (a) $C(X) = \{(x_1, x_2) \mid x_1^2 + x_2^2 \le 1\}$
   (b) $C(X) = \{x \mid x = \lambda x_1 + (1 - \lambda)\, x_2,\ 0 \le \lambda \le 1\}$

7. $\frac{1}{4}\, x_1 + \frac{1}{4}\, x_2 + \frac{1}{2}\, x_3$ where $x_1 = (0, 0)$
   $x_3 = (0, 1)$ and $x_3 = (1, 0)$ are vertices.

8. $\frac{1}{4} x_1 + \frac{1}{4} x_2 + \frac{1}{2} x_3 + \frac{1}{2} x_1 + \frac{1}{2} x_3$
   where $x_1 = (0, 0)$, $x_2 = (2, 0)$ and $x_3 = (1, 1)$.

4

Transporation

4.1 INTRODUCTION

The transportation problem (T.P. in short) is a special case of linear programming problem having simpler computation procedure with economic application. It can be regarded as a generalization of the assignment problem.

4.2 FORMULATION OF A TRANSPORTATION PROBLEM (T.P.)

[Meerut 73, 74, 79, 83. Agra 82, P.Y.D.P.C Bus. Advi U.P. 81]

The transportation problem can be described as follows.

Consider the flow of a single commodity from m number of sources (or origin) $O_1, O_2, O_3, O_4, \ldots O_m$ to n number of sinks (or destinations) $D_1, D_2, D_3, \ldots D_n$, where it is required.

Let $a_1, a_2, a_3, \ldots a_m$, be quantities produced (or stored) at the sources $O_1, O_2, O_3, \ldots O_m$, respectively and let the demand (or requirements) at the destinations $D_1, D_2, D_3, \ldots D_n$, be $b_1, b_2, b_3, \ldots b_n$, respectively. Suppose that the total quantity produced is equal to the total quantity required.

Origin (source) (Destination)

$$i.e., \sum_{i=1}^{m} a_i = \sum_{j=1}^{n} b_j \qquad ...(1)$$

In this simple model the origins are characterised in such a way that they are not connected (i.e. there are no inter flows of commodity in between them) and the same is true for the destinations. It is also assumed that there is *no backward flow* to any source point from any demand point (as shown by directed arrow-head).

Let c_{ij} = Cost of transportation of an unit from the ith source to the jth destination.

xy = The quantity transported from the ith source to the jth destination ($x_{ij} \geq 0$).

a_i = total capacity of production or storing of the commodity at the its source point, i = 1, 2 ..., m

b_j = the total amount of the demand at the jth destination point j = 1, 2, ... m

Thus, if z is the totalcost of transportation, then

$$z = \sum_{i=1}^{m} \sum_{j=1}^{n} c_{ij}\, x_{ij}$$

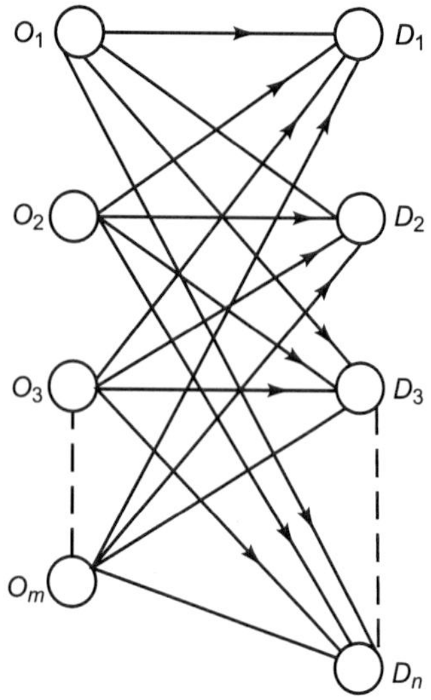

Fig. 4.1

Then the problem is to determine x_{ij} the quantity transported from the ith source to the jth destination, in such a way that the total transportation cost

$$z = \sum_{i=1}^{m} \sum_{j=1}^{n} c_{ij}\, x_{ij} \text{ is minimized.}$$

The Transportation problem as described above can be put in the following form.

Destination → Origin (sources) ↓	D_1 D_2 D_3 ... D_j ... D_n	Capacitiesof the sources
O_1	c_{11} c_{12} c_{13} ... c_{1j} ... c_{1n}	a_1
O_2	c_{21} c_{22} c_{23} ... c_{2j} ... c_{2n}	a_2
O_3	c_{31} c_{32} c_{33} ... c_{3j} ... c_{3n}	a_3
.		
.		
.		
O_i	c_{i1} c_{i2} c_{i3} ... c_{ij} ... c_{in}	a_i
.	....	.
.		
.		.
O_m	c_{m1} c_{m2} c_{m3} ... c_{mj} ... c_{mn}	a_m
Requirements →	b_1 b_2 b_3 ... b_j ... b_n	$\sum_{i=1}^{m} a_i = \sum_{j=1}^{n} b_j$

The quantity transported from the i-th sources to the j-th destination can be shown as below:

Destination → Origin (sources) ↓	D_1 D_2 D_3 ... D_j ... D_n	Capacities of the sources (supply)
O_1	x_{11} x_{12} x_{13} ... x_{1j} ... x_{1n}	a_1
O_2	x_{21} x_{22} x_{23} ... x_{2j} ... x_{2n}	a_2
O_3	x_{31} x_{32} x_{33} ... x_{3j} ... x_{3n}	a_3
.		
.		
.		
O_i	x_{i1} x_{i2} x_{i3} ... x_{ij} ... x_{in}	a_i
.	... ...	.
.		.
.	... ...	.
O_m	x_{m1} x_{m2} x_{m3} ... x_{mj} ... x_{mn}	a_m
Requirements → (Demand)	b_1 b_2 b_3 ... b_j ... b_n	$\sum_{i=1}^{m} a_i = \sum_{j=1}^{n} b_j$

These two tables can be put in a single table by writing the costs c_{ij} within one bracket ( ). Combining these two tables, we get

Destination → Origin (sources) ↓	D_1 D_2 ... D_j ... D_n	Capacities of the sources (supply)
O_1	x_{11} (c_{11}) x_{12} (c_{12})... x_{1j} (c_{1j}) ... x_{1n} (c_{1n})	a_1
O_2	x_{21} (c_{21}) x_{22} (c_{22}) ... x_{2j} (c_{2j}) ... x_{2n} (c_{2n})	a_2 .
.	....	a_3
.	....	.
O_i	x_{i1} (c_{i1}) x_{i2} (c_{i2}) ... x_{ij} (c_{ij}) ... x_{in} (c_{in})	a_i .
.	....	.
.	....	.
O_m	x_{m1} (c_{m1}) x_{m2} (c_{m2}) ... x_{mj} (c_{mj}) ... x_{mn} (c_{mn})	a_m
Requirements → (Demand)	$b_1 b_2$ ... b_j ... b_n	$\sum_{i=1}^{m} a_i = \sum_{j=1}^{n} b_j$

The above problem of transportation may be stated mathematically as

Minimize $$z = \sum_{i=1}^{m} \sum_{j=1}^{n} c_{ij}\, x_{ij} \qquad \text{...(1)}$$

subject to the constraints

$$\sum_{j=1}^{n} x_{ij} = a_i,\ i = 1, 2, 3, m$$

$$\sum_{i=1}^{m} x_{ij} = b_j,\ j = 1, 2, 3, n$$

$$\sum_{i=1}^{m} a_i = \sum_{j=1}^{n} b_j \ \ldots\ldots.. \qquad \text{...(2)}$$

and $x_{ij} \geq 0$, for all $i = 1, 2, 3 \ldots m$

$j = 1, 2, 3, \ldots n$

Balanced Transportation Problem—

If the total amount of supply = the total amount of demand

i.e. $$\sum_{i=1}^{m} a_i = \sum_{j=1}^{n} b_j$$

If $\sum_{i=1}^{m} a_i \neq \sum_{j=1}^{n} b_j$, the problem is called Unbalanced Transportation problem.

4.3 SOME USEFUL DEFINITIONS

Following are some useful terms which are frequently used in the discussion of the transportation problem.

(i) A feasible solution:— A feasible solution of a transportation problem is a set of non-negative individual allocations ($x_{ij} \geq 0$) which satisfies the row and column sum restrictions. That is constraints (2) of Article 4.2).

(ii) Rim Requirements :— The availability and the demand constraints are called Rim Requirements. (i.e. constraints 2 of Art. 4.2)

(iii) Basic feasible solution :— A feasible solution of m by n transportation problem is said to be a Basic Feasible Solution if the total number of positive allocations x_{ij} is exactly equal to $m + n - 1$ (that is one less than the sum of number of rows and columns).

(iv) Optimal Solution :— A feasible solution not necessarily basic is said to be optimal if it minimizes the total transportation cost.

(v) Non-degenerate Basic feasible solution :— (Agra 84)

A feasible solution of m by n transportation problem is said to be *non-degenerate basic feasible solution* if (a) total number of positive allocations is exactly equal to $(m + n - 1)$ and (b) these allocations are at independent positions.

This can be put in the following way—

If a feasible solution involves exactly $(m + n - 1)$ independent individual positive allocations, then it is known a *Non-degenerate basic feasible solution*, otherwise it is said to be *degenerate basic feasible solution.*

(vi) Independent position of allocation :—

By independent positions of the allocations we mean that it is always impossible to form any closed circuit (loop by joining these allocations by horizontal and vertical lines only). This can be understood by the following tables in which the positions of allocations are denoted by ○.

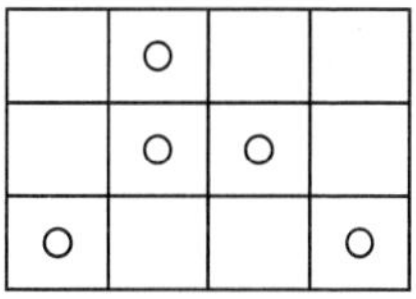

Independent position

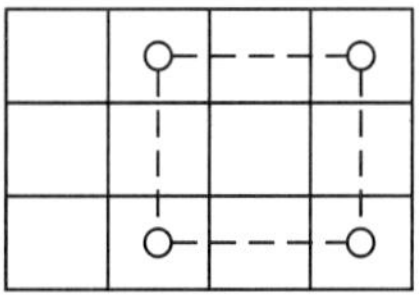

Non-independent position

4.4. SOLUTION OF A TRANSPORTATION PROBLEM

To obtain the solution (optimal) of a transportation problem we mean the following :—

(i) to obtain an initial basis feasible solution.

(ii) to find an optimal solution by making successive improvements in the basic feasible solution obtained in (i) until no further decrease in the transportation cost is possible.

4.5 BASIC FEASIBLE SOLUTION OF A BALANCED TRANSPORTATION PROBLEM

There are many interesting methods available for finding the basic feasible solution of a balance transportation problem. Here, we shall discuss a few of them.

(i) North-West Corner Rule.
(ii) Lowest Cost Entry method (or Method of Matrix minimum)
(iii) Vogel approximation method or V. A. M.
(iv) Hungarian Method.

Before going into the details of these methods some important theorems regarding these will be proved below.

Some Important Theorems.

4.6 THEOREM I

[V. B. U. pass III Hons. 96, 98]

The condition $\sum_{i=1}^{m} a_i = \sum_{j=1}^{n} b_j$ (i. e. total supply = total demand) is both necessary and sufficient condition for the existence of the feasible solution of transportation problem:—

$$\text{Minimum} \quad z = \sum_{i=1}^{m} \sum_{j=1}^{n} c_{ij} x_{ij} \quad \ldots(1)$$

subject to restrictions

$$\sum_{j=1}^{n} x_{ij} = a_i, \; i = 1, 2, 3 \ldots m \quad \ldots(2)$$

$$\sum_{i=1}^{m} x_{ij} = b_j, \; j = 1, 2, \ldots n \quad \ldots(3)$$

$$x_{ij} \geq 0 \quad \ldots(4)$$

This can be put in the following way—
If all the sources are emptied and all the destinations are filled, show that $\sum a_i = \sum b_j$ is a necessary and sufficient condition for the existence of a feasible solution to a transportation problem.

[Delhi O. R. 66, Sardar Patel University M. Stat, 75]

Proof : The condition is necessary.

Suppose there exists a feasible solution $x_{ij} = x^*_{ij}$ to the transportation problem (1) to (4). From (2) we have, for i = 1, 2, 3, ...m

$$a_i = \sum_{j=1}^{n} x_{ij}^*$$

$$\text{i.e. } \sum_{i=1}^{m} a_i = \sum_{i=1}^{m} \sum_{j=1}^{n} x_{ij}^*$$

$$= \sum_{j=1}^{n} \sum_{i=1}^{m} x_{ij}^* = \sum_{j=1}^{n} b_j \text{, by (3)}$$

This prove the result.
The condition is sufficient.

$$\text{Let} \quad \sum_{i=1}^{n} a_i = \sum_{j=1}^{n} b_j = M \text{ (say)} \qquad ...(5)$$

Let $\beta_i \geq 0$ be any real number such that
$x_{ij} = \beta_i \, b_j$ for all i, j.
Hence, β_i is given by

$$a_i = \sum_{j=1}^{n} x_{ij} = \sum_{j=1}^{n} \beta_i . b_j \, , \forall \, i$$

$$= \beta_i \sum_{j=1}^{n} b_j$$

$$= \beta_i \, M \text{, from (5)}$$

$$\text{or,} \quad \beta_i = \frac{a_i}{M} \text{, i = 1, 2, ...} m$$

Thus one of the possible solution of $x_{ij} = \beta_i b_j = \dfrac{a_i b_j}{M}$, for all i, j; obviously, such $x_{ij} \geq 0$ because $a_i \geq 0$, $b_j \geq 0$ for all i, j.
Hence, a feasible solution of the Transportation problem exists.

4.7 THEOREM 2

The transportation problem remains unchanged with respect to the optimal solution if all elements of the cost matrix are increased or decreased by the same amount.

Proof : Let $c'_{ij} = c_{ij} + k$ where k may be negative or positive.

Now,
$$z = \sum_{i=1}^{m} \sum_{j=1}^{n} c_{ij}\, x_{ij}$$

$$= \sum_{i=1}^{m} \sum_{j=1}^{n} (c'_{ij} - k)\, x_{ij}$$

$$= \sum_{i=1}^{m} \sum_{j=1}^{n} c'_{ij} - k \sum_{i=1}^{m} \sum_{j=1}^{n} x_{ij}$$

$$= \sum_{i=1}^{m} \sum_{j=1}^{n} c'_{ij} - kM$$

Thus we see that in the above expression only the first term is a variable quantity and the 2nd term kM is a constant. Since addition or subtraction of a constant term with the objective function z does not change the optimal values if any of x_{ij}'s, the theorem follows easily.

4.8 THEOREM 3

In the balanced transportation problem

$$\text{Min. } z = \sum_{i=1}^{m} \sum_{j=1}^{n} c_{ij}\, x_{ij} \qquad ...(1)$$

Subject to the restriction

$$\sum_{j=1}^{n} x_{ij} = a_i\,,\ i = 1, 2, 3 \,....m \qquad ...(ii)$$

$$\sum_{i=1}^{m} x_{ij} = b_j\,,\ j = 1, 2, 3 \,...n \qquad ...(iii)$$

$$x_{ij} \geq O \text{ and } \sum_{i=1}^{m} a_i = \sum_{j=1}^{n} b_j \qquad ...(iv)$$

the number of non-degenerate basic feasible solutions almost $m + n - 1$

Proof : We see that the above transportation problems is a linear programming problem with $(m + n)$ constraints. But only $(m + n - 1)$ of them are the independent. This means that one of them is redundant. For this add all the row equation in (ii) and from this subtract the sum of the first $(n - 1)$ column equations in (iii) to get the last column equation $(j = n)$ in (iii).

Adding all equations in (ii) gives us

$$\sum_{i=1}^{m} \sum_{j=1}^{n} x_{ij} = \sum_{i=1}^{n} a_i \qquad \text{...(v)}$$

Adding first $(n-1)$ equations in (iii), we get

$$= \sum_{j=1}^{n-1} \sum_{i=1}^{m} x_{ij} \sum_{j=1}^{n-1} b_j \qquad \text{... (vi)}$$

Subtracting (vi) from (v) we get

$$\sum_{i=1}^{m} \left[\sum_{j=1}^{n} x_{ij} - \sum_{j=1}^{n-1} x_{ij} \right] = \sum_{i=1}^{n} a_i - \sum_{j=1}^{n-1} b_j$$

$$= \sum_{j=1}^{n} b_i - \sum_{j=1}^{n-1} b_i = b_n, \text{ from (iv)}$$

Hence, $\sum_{i=1}^{n} x_{ij} = b_n$ which is the last equation in (iii)

This proves the result.

4.9 THEOREM 4

Difference between a transportation and an assignment problem.

(R. U. Condensed Course Hons. 89, Meerut 70, 73, 75, 89; Raj. 79, 85)

An assignment problem is a special case of the transportation problem in which

(i) $m = n$

(ii) all the a_i and b_i are unity

(iii) each x_{ij} is limited to one of the two values 0 and 1.

In such situations exactly n of x_{ij} can be non-zero, one in each row of the array and one in each column, showing that only one source can be allocated to each destination.

4.10 THEOREM 5

The transportation problem remains unchanged with respect to the optimal solution if the elements in any particular row or column of the cost matrix are increased by the same amount.

Proof : The objective function in a transportation problem is

$$z = \sum_{i=1}^{m} \sum_{j=1}^{n} c_{ij}\, x_{ij}$$

Let $c_{ij} = c'_{ij} + p_i + q_j$, where p_i and q_j are arbitrary constants.

Then

$$z = \sum_{i=1}^{m} \sum_{j=1}^{n} (c'_{ij} + p_i + q_j)\, x_{ij}$$

$$= \sum_{i=1}^{m} \sum_{j=1}^{n} c'_{ij} x_{ij} + \sum_{i=1}^{m} \sum_{j=1}^{n} p_i\, x_{ij} + \sum_{i=1}^{m} \sum_{j=1}^{n} q_j x_{ij}$$

$$= \sum_{i=1}^{m} \sum_{j=1}^{n} c'_{ij} x_{ij} + \sum_{i=1}^{m} p_i \sum_{j=1}^{n} x_{ij} + \sum_{j=1}^{n} q_j \sum_{i=1}^{m} x_{ij}$$

$$= \sum_{i=1}^{m} \sum_{j=1}^{n} c'_{ij} x_{ij} + \sum_{i=1}^{m} a_i\, p_i + \sum_{j=1}^{n} b_j\, q_j$$

Since the last two terms are constants (second term is a sum over all i. and the last term is a sum over all j's), and the only variable is the first one. therefore, the theorem follows.

4.11 NORTH-WEST CORNER RULE

(C. A. Final May 87, Delhi 71, Raj, 74)

Here, we shall use the term "cell" (i, j). We mean by (i, j) the element at the intersection of the ith row and jth column.

By this rule we assign a set of allocations in the cells so that the row totals and column totals will be indicated before each ( ) as follows.

(i) Start with cell (1, 1) from the left hand (or North-West corner i. e. the topmost left corner) and assign it maximum possible amount, i. e., x_{11} = min. (a_1, b_1)

(ii) Then move to the right hand cell (1, 2) if there is still any available quantity left, otherwise move to the down cell (2, 1) and assign it maximum possible amount.

(iii) Repeat the step (ii) again and continue until all the available quantity is exhausted.

This method can be well explained by solving the following numerical example.

Example 1. Determine an initial basic feasible solution to the following Transportation Problem by using North-West corner Rule.

		Destination					Supply
		D_1	D_2	D_3	D_4	D_5	
	O_1	(2)	(11)	(10)	(3)	(7)	4
Origin	O_2	(1)	(4)	(7)	(2)	(1)	8
	O_3	(3)	(9)	(4)	(8)	(12)	9
Demand		3	3	4	5	6	

(Delhi 1973)

Solution: We start with the top most left corner and assign to it the maximum possible amount $x_{11} = 3$ (since minimum of $a_1 = 4$ and $b_1 = 3$ is 3), assignment 3 is shown in the top most left corner by

		D_1	D_2	D_3	D_4	D_5	*Supply*
	O_1	3 (2)	1 (11)				$4 = a_1$
Origin	O_2		2 (4)	4 (7)	2 (2)		$8 = a_2$
	O_3				3 (8)	6 (12)	$9 = a_3$
Demand		$3 = b_1$	$3 = b_2$	$4 = b_3$	$5 = b_4$	$6 = b_5$	

Putting 3 before ( ); and then the demand at D_1 is satisfied. However, still the capacity at the origin D_1 is not exhausted, and thus we assign x_{12} = minimum [4 – 3, 3] = 1 (that is, we move towards the right). Now, at this stage the capacity 4 units of O_1 is exhausted but not the demand of d_2. Hence, we move down and assign x_{23} = minimum [8, 3 – 1] = 2.

Now, we find that the capacity at O_2 is not exhausted. So we move towards the right and assign x_{22} min. [8 – 2, 4] = 4. However, still the capacity at O_2 is not exhausted. So again we move towards right and assign x_{24} = mini. [8 – 2 – 4, 2] = 2. Now, at this stage the capacity $a_2 = 8$ of O_2 is exhausted but the demand of D_4 is not satisfied. Hence, we move down and assign x_{34} = min. [9, 5–2] = 3. However, still the capacity at O_3 is not exhausted and so we move towards right and assign x_{35} = min. [9 – 3, 6] = 6. At this stage the capacity of O_3 is exhausted.

This completes the allocations and the resulting feasible solutions are shown in the table.

The unoccupied cells are clearly assigned with zeros values. On multiplying each individual allocation by its corresponding unit in ( ), and adding the total transportation cost to this feasible solutions is

$$2 \times 3 + 11 \times 1 + 4 \times 2 + 7 \times 4 + 2 \times 2 + 8 \times 3 + 12 \times 6 = 153 \text{ units.}$$

Note 1. North-West corner rule is very simple but it does not take care of the magnitude of values of c_{ij}'s and thus normally does not have coverage to the optimal solution.

2. In the above problem, $m = 3$, $n = 5$. Hence the basic feasible solution $x_{11} = 3$, $x_{12} = 1$, $x_{22} = 2$, $x_{23} = 4$, $x_{24} = 2$, $x_{34} = 3$, $x_{35} = 6$ is *non-degenerate*, because the total number of non-zeros x_{ij}'s are 3 + 5 – 1 = 7

Example 2. Determine the initial basic feasible solution to the following transportation problem, using North-West Corner Rule.

		Destination D_1	D_2	D_3		Supply
	O_1	(2)	(7)	(4)	5	
Origin	O_2	(3)	(3)	(1)	8	
	O_3	(5)	(4)	(7)	7	
	O_4	(1)	(6)	(2)	14	
Demand		7	9	18	34	

Solution : Costs are shown in ( ). Assignment (or allocation) will be put ( ).

(i) Start from upper left hand North-West Corner and assign to the cell (1, 1) the maximum possible amount 5 (since maximum of $a_1 = 5$ and $b_1 = 7$ is 5)

Destination		D_1	D_2	D_3	Supply
	O_1	5(2)			$5 = a_1$
Origin	O_2	2(3)	6(3)		$8 = a_2$
	O_3		3(4)	4(7)	$7 = a_3$
	O_4			14(2)	$14 = a_4$
Demand →		$7 = b_1$	$9 = b_2$	$18 = b_3$	

(ii) Since there is no amount left available at the source O_1, so we move downward to the cell (2, 1) in place of moving to right and allot it maximum possible amount since the column D_1 still need the amount 2 and the amount 8 is available in row O_2 so we allocate the amount 2 and the amount 8 is available in row O_2 so we allocate the maximum amount 2 to this cell (2, 1), Now, the allocation for the column D_1 is complete, so we move to the right of this cell. Since the amount 6 is still available in row 2 and amount 9 required in column D_2, so we assign the maximum amount 6 in this cell (2, 2). This assignment for row O_2, is complete, so we move downward to the cell (3, 2). Since the amount 3 is still required in column D_2 and amount 7 is available in row a_3, so we assign the maximum amount 3 in this cell (3, 2). Thus allocation for column D_2 is complete, so we move to the right of this cell. Since the amount 4 is still available in row O_3 and the amount 18 is needed in column D_3, so we assign the maximum amount 4 to the downward cell (3, 3). Thus the requirement of the row O_3 is complete, so we move to the downward cell (4, 3). Since the amount 14, is still available in now O_3 and an equal amount 14 is required in row O_4, so we allocate this amount 14 to this cell (4, 3). This completes the allocations and the resulting feasible solution (allocations) are shown in the table.

On multiplying each individual allocation by its corresponding unit in ( ), and adding, the total transportation cost to this feasible solution is

$$= 5 \times 2 + 2 \times 3 + 6 \times 3 + 3 \times 4 + 4 \times 7 + 14 \times 2 = \text{Rs. } 102.$$

Note (i) In this North-West Corner rule we move to the right or down so no loop (circuit) will be formed by drawing horizontal and vertical lines to the allocations.

(ii) Here $m = 4$, $n = 3$. The number of basic feasible solution is $6(x_{11} = 5, x_{21} = 2, x_{22} = 6, x_{32} = 3, x_{33} = 4, x_{43} = 14)$

Hence, it is non-degenerate because the total number of non zero x_{ii}'s is 6.

$$= (m + n - 1 = 4 + 3 - 1 = 6)$$

4.12 METHOD OF MATRIX MINIMA (OR LOWEST COST ENTRY METHOD)

The costs are written within bracket ( ). Following steps are adopted.

(i) Choose the cell with the lowest cost and assign it as much as possible.
If such cell of the lowest cost is not unique, select any one of these cells.

(ii) Again, examine the cost matrix and select the cell with the lowest cost (the cell in which the allocation has been made is not considered). Allocate there as much as possible.

(iii) Continue this procedure until all the available quantity is exhausted.

This method is explained by the following numerical example.

Example 3. Solve the following transportation problem.

	Destinatio				
	D_1	D_2	D_3		Supply
Origin	O_1	*(2)*	*(7)*	*(4)*	5
	O_2	*(3)*	*(3)*	*(1)*	8
	O_3	*(5)*	*(4)*	*(7)*	7
	O_4	*(1)*	*(6)*	*(2)*	14
Demand		7	9	18	

Solution : We see that the lowest cost 1 is at (2, 3) and (4, 1), choose any one of them, say cell (2, 3) and allot the maximum amount 8 to this cell. Leaving this cell, we find other lowest 1 is at

(4, 1) where the maximum amount 7 is allotted. Now, the next lowest cost 2 is in column D_1 at (1, 1) and in column D_3 at (4, 4). But column D_1 already got its requirement. So we leave it and allocate maximum possible amount 7 to (4, 4). The availability of row O_4 is exhausted (out of a1 = 14), amount 7 goes to (4, 1) and goes to (4, 4). Next lowest cost is 3 which is in row O_2 where availability $a_2 = 8$ has, already exhausted. After this, we see that the next lowest cost is 4 at (1, 3) and (3, 2) we allocate only 3 to (1,) because the requirement of D_3 is 18 out of which 15 (8 and 7) already allocated to D_3 and assign 7 to (3, 2). A allocation of D_1 and D_3 are complete. The requirement of the column D_2 will be completed by allocating 2 to (1, 2) out of the remaining availability of row O_1.

D_1	D_2	D_3	
(2)	2(7)	3(4)	$5 = a_1$
(3)	(3)	8(1)	$8 = a_2$
(5)	7(4)	(7)	$7 = a_3$
7(1)	(6)	7(2)	$14 = a_4$
$b_1 = 7$	$b_2 = 9$	$b_3 = 18$	

The required feasible solution is shown in the table. The total transportation cost to this is

$$= 2 \times 7 + 3 \times 4 + 8 \times 1 + 7 \times 4 + 7 \times 1 + 7 \times 2 = 83$$

Example 4. Solve the following transportation problem by matrix minima method :

Destination →		D_1	D_2	D_3	D_4	D_5	Supply
	O_1	(2)	(11)	(10)	(3)	(7)	4
Origin	O_2	(1)	(4)	(7)	(2)	(1)	8
	O_3	(3)	(9)	(4)	(8)	(12)	9
Demand		3	3	4	5	6	

The lowest cost 1 is at (2, 1) and (2, 5) choose any one of them, say, cell (2, 5) and allocate the maximum 6 to this cell leaving this cell, we find that other lowest cost 1 is (2, 1) where the maximum possible amount 2 is assigned. Continuing in this way we get the required feasible solution shown in the table.

	D_1	D_2	D_3	D_4	D_5	
O_1	1(2)	(11)	(10)	3(3)	(7)	4
O_2	2(1)	(4)	(7)	(2)	6(7)	8
O_3	(3)	3(9)	4(4)	2(8)	(12)	9
	3	3	4	5	6	

The value of transportation cost is

$$= 1 \times 2 + 3 \times 3 + 2 \times 1 + 6 \times 1 + 3 \times 9 + 4 \times 4 + 2 \times 8$$

$$= 78 \text{ units}$$

Note :—

The above problem can be solved in detail as follows.

In the problem the minimum value of cost (c_{ij}) occur at the cells (2, 1) and (2, 5). Choose any of them. Select (2, 5) cell and allocate $x_{35} = 6$.

See the adjoining table:

	D_1	D_2	D_3	D_4	D_5	
O_1	(2)	(11)	(10)	(3)	(7)	4
O_2	(1)	(4)	(7)	(2)	6(1)	8
O_3	(3)	(9)	(4)	(8)	(12)	9
	3	3	4	5	6	

Since the demand at the column D_5 is satisfied, we strike off the 5 the column form the cost matrix and obtain the following matrix.

	D_1	D_2	D_3	D_4	
O_1	(2)	(11)	(10)	(3)	4
O_2	2(1)	(4)	(7)	(2)	2
O_3	(3)	(9)	(4)	(8)	9
	1	3	4	5	

Now, here the c_{ij} in (2, 1) cell is minimum. Hence assign $x_{21} = 2$ and the 2nd row is crossed off since the cost 2 at the cell (1, 1) is now minimum we allocate 1 there and the first column is satisfied. The reduced cost matrix is now

	D_1	D_2	D_3	D_4	
O_1	(12)	(11)	(10)	(3)	3
O_3	(3)	(9)	(4)	(8)	9
	1	3	4	5	

The final allocation is follows

	D_1	D_2	D_3	
O_1	(11)	(10)	3(3)	3
O_2	3(9)	(4)	2(8)	9
	3	4	5	

Now the full assignment can be shown in the following matrix.

	D_1	D_2	D_3	D_4	D_5	
O_1	1(2)	(11)	(10)	3(3)	(7)	4
O_2	2(10	(4)	(7)	(2)	6(1)	8
O_3	(3)	3(9)	4(4)	2(8)	(12)	9
	3	3	4	5	6	

The total transportation cost is

$$Z = 2 \times 1 + 3 \times 3 + 1 \times 2 + 1 \times 6 + 9 \times 3 + 4 \times 4 + 8 \times 2 = 78 \text{ units.}$$

4.13 VOGEL'S APPROXIMATION METHOD (UNIT COST PENALTY METHOD)

(Rohil Khand 85, CA 87, P. G. Dept. (Bus. Adm.) 81)

In this method the following steps are adopted.

(i) Write the differences of the smallest and the second smallest costs in each column below the corresponding column.

(ii) Write the differences of the smallest and the second smallest cost of each row to the right of the corresponding row. These individual differences are regarded as a *penalty* for making assignments in the second lowest cost cell in place of the lowest cost cell in each row (or column).

(iii) Select the row or column for which the penalty is the largest and allocate the maximum possible amount to the cell with lowest cost in that particular row or column. If there are more than one largest penalty rows or columns, then select any one of them.

(iv) Cross (or leave) or column in which the requirement (or demand) has been satisfied and then construct the reduced matrix.

(v) Continue this process on the reduced matrices till all the allocations have been exhausted (made). Consider the following example for this method.

Example 5. Solve of the following transportation problem, using Vogel's approximation method.

		Destination D_1	D_2	D_3	Supply
Origin	O_1	(2)	(7)	(4)	5
	O_2	(3)	(3)	(1)	8
	O_3	(5)	(4)	(7)	7
	O_4	(1)	(6)	(2)	14
Demand		7	9	18	

(Dibrugarh M. Sc. Stat. 76, Meerut O R, 70, 77, M.Com. 77, VBU Part III Hons 99)

Solution : In the first column the smallest cost is 1 and second smallest cost is 2. Their difference is 2 – 1 = 1 which has been written below the column D_1. In the same way differences of the smallest and second smallest of columns D_2 and D_3 are shown below them.

	D_1	D_2	D_3	Available		Penalties
O_1	5(2)	(7)	(4)	5	←	(4 – 2 = 2)
O_2	(3)	(3)	(1)	8		(3 – 1 = 2)
O_3	(5)	(4)	(7)	7		(5 – 4 = 1)
O_4	(1)	(6)	(2)	14		(2 – 1 = 1)
Demand	7	9	18			
Penalties	(2 – 1 = 1)	(4 – 3 = 1)	(2 – 1 = 1)			

Similarly; the differences in the smallest cost and second smallest of each row are shown right of the corresponding row.

Since the maximum penalty 2 occurs in rows O_1 and O_2, select any one of them, say row O_1. Then allocate the maximum possible amount 5 to the cell (1, 1) with lowest cost and cross this row O_1. The first reduced matrix after striking off row O_1 with remaining demands and available is as below. See the amount 5 has been allotted to column D_1. So the amount still needed to column D_1 is 2.

	D_1	D_2	D_3	Available	Penalties
O_2	(3)	(3)	8(1)	8	← (3 – 1 = 2)
O_3	(5)	(4)	(7)	7	(5 – 4 = 1)
O_4	(1)	(6)	(2)	14	(2 – 1 = 1)
Demand	2	9	18		
Penalty	(3 – 1 = 2)	(4 – 3 = 1)	(2 – 1 = 1)		

For this, penalties are calculated on the same basis. Since the maximum penalty 2 is associated with row O_2 therefore, the maximum possible amount 8 is assigned to the cell (1, 3) with lowest cost in this row O_2.

Now, strike off the row O_2 and form the second reduced matrix after leaving row O_2 with remaining demands and availables; The reduced matrix is as below.

	D_1	D_2	D_3	Available	Penalties
O_3	(5)	(4)	(7)	7	(5 – 4 = 1)
O_4	(1)	(6)	10(2)	14	(2 – 1 = 1)
Demand	2	9	10		
Penalties	(5 – 1 = 4)	(6 – 4 = 2)	(7 – 2 – 5)		

See the maximum penalty 5 is associated with column D_3 in which the lowest cost 2 occurs at (2, 3). So allocate the maximum possible amount 10 to this cell (2, 3).

In the same way the next reduced matrix after striking off column D_3 with remaining demands and available is as below.

	D_1	D_2	Available	Penalties
O_3	(5)	(7)	7	1
O_4	2(1)	2(6)	4	5
Demand	2	9		
Penalties	4	2		

See the maximum penalty 5 occurs in row O_4. So allocate the maximum amount 2 to the cell (2, 1) with lowest cost in this row. The remaining amount 2 in the row O_4 is allocated the cell (2, 2) and then we allote the remaining amount 7 to the cell (1, 2) with smallest cost after striking off the row O_4.

The full allocation is thus given by

	D_1	D_2	D_3	Available
O_1	5(2)	(7)	(4)	5
O_2	(3)	(3)	8(1)	8
O_3	(5)	7(4)	(7)	7
O_4	2(1)	2(6)	10(2)	14
Demand	7	9	18	

The total transportation cost is

$$Z = 5 \times 2 + 8 \times 1 + 7 \times 4 + 2 \times 1 + 2 \times 6 + 10 \times 2 = 80$$

4.14 THE HUNGRIAN METHOD

Following steps are employed to solve the transportation problem by this method:

(i) Subtract the smallest element of each row from each element of that row.

(ii) Locate those columns in which there is no zero after employing first step. Subtract the minimum element of such non - zero columns from that column.

(iii) Since the changed cost matrix is equivalent to the original one allocate if possible $(m + n - 1)$ number x_{ij}'s in the (i, j) th cell where ever zero appears.

(iv) If the allocation does not give $(m + n - 1)$ number of non - zero x_{ij}'s, then repeat steps (i) and (ii).

Below we shall illustrate this method with example.

Example 6. Solve the following transportation problem by Hunge-rain method.

	D_1	D_2	D_3	D_4	D_5	
O_1	2	11	10	3	7	4
O_2	1	4	7	2	1	8
O_3	3	9	4	8	12	9
Requirement	3	3	4	5	6	

Solution : Minimum values in the 1st, 2nd 3rd rows are 2, 1, 3 respectively. Subtract these from their respective rows.

The matrix obtained is

	D_1	D_2	D_3	D_4	D_5	
O_1	2—2	11—2	10—2	3—2	7—2	4
O_2	1—1	4—1	7—1	2—1	1—1	8
O_3	3—3	9—3	4—3	8—3	12—3	9
	3	3	4	5	6	

This is equivalent to

	D_1	D_2	D_3	D_4	D_5	
O_1	0	9	8	1	5	4
O_2	0	3	6	1	0	8
O_3	0	6	1	5	9	9
	3	3	4	5	1	

Now, the non-zero columns are D_2, D_3, D_4 only. In these column there are no zero subtract the smallest element 3 of D_2 from its each element. The elements of D_2 become 9 — 3, 3 — 3, 6 — 3, i.e. 6, 0, 3.

Similarly, the elements of D_3 are 8 1, 6 — 1, 1 — 1, i.e. 7, 5, 0 and those of D_4 are 1—1, 1—1, 5—1. The changed matrix is

	D_1	D_2	D_3	D_4	D_5	
O_1	0	6	7	0	5	4
O_2	0	0	5	0	0	8
O_3	0	3	0	4	9	9
	3	3	4	5	6	

For allocation we may start from (2, 5) cell with 6, then to (2, 4) with 2, then to (1, 4) with 3, then to (1, 1) with 1 to satisfy the rim requirements row requirements 4, 8, 9; column requirements (3,

3, 4, 5, 6). We fail to use zeros at (2, 1) and (2, 2). We then allocate 2 to (3, 1) and 4 to (3, 3) Lastly, we are forced to use (3, 2) with 3 to satisfy the rim requirements. Thus the final allocation would be

	D_1	D_2	D_3	D_4	D_5	
O_1	1 (0)	6	7	3 (0)	5	4
O_2	0	0	5	2 (0)	6 (0)	8
O_3	2 (0)	3 (3)	4 (0)	4	9	9
	3	3	4	5	6	

The total cost is given by (consulting the original cost matrix in the occupied cell) is

$Z = 2 \times 1 + 3 \times 3 + 2 \times 2 + 1 \times 6 + 3 \times 2 + 9 \times 3 + 4 \times 4 = 70$

Note — If we start with $x_{11} = 3$, then we get $x_{11} = 3, x_{14} = 1, x_{22} = 3, x_{24} = 4, x_{25} = 1, x_{33} = 4, x_{35} = 5$.

$Z = 3 \times 2 + 1 \times 3 + 3 \times 4 + 4 \times 2 + 1 \times 1 + 4 \times 4 + 5 \times 12 = 106$ units.

We see that in this location the cost is not improved.

Example 7. Solve the following transportation problem:

To

		W_1	W_2	W_3	W_4	Available
	F_1	(1)	(2)	(1)	(4)	30
	F_2	(3)	(3)	(2)	(1)	50
From	F_3	(4)	(2)	(5)	(9)	20
Demand		20	40	30	10	

Bombay B. Sc. (Stat) 76, Kurukshetra 69, 77, Delhi 68, Meerut M. Sec 69]

Solution : We solve this problem by

(i) North-West corner method

(ii) Vogel's approximation method

(iii) Matrix minima.

(i) North-West Corner rule.

In this method we start form the top most left corner i.e. (1, 1) cell and allocate there maximum possible amount. Thus 20 which is Min. ($a_1 = 30, b_1 = 20$) can be allocated in this. We cannot send more than 20. After this allocation 10 units are still available at F_1. So we move to the next cell of the first row and assign as much as possible. Here we cannot allocate more than 10 because only 10 units are available at F_1.

To

	W_1	W_2	W_3	W_4	$a_j \downarrow$
F_1	20 (1)		10 (1)		$30 = a_1$
F_2		30 (3)	20 (2)		$50 = a_2$
F_3			10 (5)	10 (9)	$20 = a_3$
	20 $= b_1$	40 $= b_2$	30 $= b_3$	10 $= b_4$	

But still demand of column W_2 is not satisfied. So we allocate more in the 2nd column.

We move to the next cell of the second column i.e. to the cell (2, 2). We assign as much as possible. To this cell we cannot allocate more than 30 otherwise the sum of second column will exceed 40. Still there are 20 units available at F_2. We can send there 20units to the next destination i.e. W_3. Thus we allocate 20 units to (2, 3) cell. But still 20 units are needed to W_3. We allocate 10 units to (3, 3). The remaining 10 units available at F_3 are allotted to (3, 4) the resulting feasible solution is shown in the matrix.

The transportation cost corresponding to this feasible solution is

$$Z = 20 \times 1 + 10 \times 1 + 30 \times 3 + 20 \times 2 + 5 \times 10 + 9 \times 10 = 300$$

(ii) By Vogel's approximation method:

Write the difference of the smallest and the second smallest costs in each column and write it below the column. In the same way write the difference infront of corresponding row. We get the following table.

	W_1	W_2	W_3	W_4	Available	Penalty
F_1	(1)	(2)	(1)	(4)	30	1
F_2	(3)	(3)	(2)	10 (1)	50	1
F_3	(4)	(2)	(5)	(9)	20	2
Requirement	20	40	30	10		
Penalties	2	1	1	3 ↑		

The maximum penalty is 3 where we have marked (↑). It signifies that if we depart for the minimum cost cell in this column W_4, then the cost will increase at least by 3. Thus if we don't assign to this cell of minimum cost i.e. cell (2, 4) as much as possible. We can allocate here atmost 10. Since the demand of this column is satisfied, it is not required to be written. So we omit it. The reduced cost matrix is

	W_1	W_2	W_3	Available	Penalty
F_1	(1)	(2)	(1)	30	(1)
F_2	(3)	(3)	(2)	40	(1)
F_3	(4)	20 (2)	(5)	20	←(2)
Requirement	20	40	30		
Penalties	(2)	(1)	(1)		

Here, the greatest penalty is 2. We may assign in the first column or in the 3rd row. Here we allocate 20 units to the minimum cost cell (3, 2). Since the 3rd row has no more available, we omit it and write the reduced matrix as below.

	W_1	W_2	W_3	Available	Penalty
F_1	20(1)	(2)	(1)	30	(1)
F_2	(3)	(3)	(2)	40	(1)
Requirement	20	20	30		
Penalties	(2)	(1)	(1)		

↑

Now, we allocate in the first column. We allocated 0 to (1, 1). Since requirement of column 1 is satisfied, we omit the column and write the reduced matrix as follows—

	W_2	W_3	Available	Penalty
F_1	(2)	10 (1)	10	1
F_2	20 (3)	20 (2)	40	1
Requirement	20	30		
Penalties	1	1		

We assign 10 to (1, 2) of this matrix and hence to meet the demand of W_3, 20 units are allocated to (2, 2) of this matrix. The remaining 20 units of F_2, are allocated to (2, 1) of this matrix.

Thus the final feasible solution is

	W_1	W_2	W_3	W_4	Available
F_1	20 (1)	(2)	10 (1)	(4)	30
F_2	(3)	20 (3)	20 (2)	10 (1)	50
F_3	(4)	20 (2)	(5)	(9)	20
Requirement	20	40	30	10	

The transportation cost corresponding to this feasible solution is

$$Z = 20 \times 1 + 10 \times 1 + 20 \times 3 + 20 \times 2 + 10 \times 1 + 20 \times 2 = 180$$

(iii) By matrix minima (Lowest Cost Entry Method)

	W_1	W_2	W_3	W_4	Available
F_1	20 (1)	(2)	10 (1)	(4)	30
F_2	(3)	20 (3)	20 (2)	10 (1)	50
F_3	(4)	20 (2)	(5)	(9)	20
Demand	20	40	30	10	

The lowest cost 1 is at (1, 1) (1, 3) and (2, 4). Allocate maximum possible amount to these cells. Allocate 20 to (1, 1), 10 to (1, 3) and 10 to (2, 4). The requirements of columns 1 and 4 are stisfied. Second lowest cost 2 is at (1, 2), (2, 3) and (3, 2). Allocate 20 to (2, 3) 20 to (3, 2), (2, 2) will not get allotment as availablity 30 of F_1 has already exhausted. Next lowest cost 3 is at (2, 2) where maximum possible allocation 20 can be made. The final allocations are shown in the above matrix. The trnasportation cost is

$$Z = 20 \text{ x } 1 + 10 \text{ x } 1 + 20 \text{ x } 3 + 20 \text{ x } 2 + 10 \text{ x } 1 + 20 \text{ x } 2 = 180$$

Note— Comparing the results of the above problem obtained by three different methods we conclude that the Vogel's approximation method or Matrix maximum method gives much better solution than the north-west corner rule.

Unbalanced Transportation Problem

If in a transportation problem

$$\sum_{i=1}^{m} a_i \neq \sum_{j=1}^{n} b_j$$

then the problem is known as *Unbalanced* transportation problem.

The unbalanced transportation problem can be solved in the following way :

(i) Modify the unbalanced problem into a balanced problem. This can be done by introducing a fictious (dummy) source or destinations which will provide the surplus, supply or demand respectively.

(ii) The costs of transporting an unit item from fictious source are taken zero.

(iii) Similarly, the costs of transporting an unit to the fictious destination is also taken zero.

This can be well explained by the following example.

Example 8. Solve the following transportation problem—

	W_1	W_2	W_3	Available
F_1	(30)	(20)	(10)	800
F_2	(5)	(15)	(25)	500
Requirement	300	300	400	

Solution : We see that in this problem the total available exceeds the total demand. We consider a fictious destination requiring

$$\sum_{i=1}^{m} a_i - \sum_{j=1}^{n} b_j = 1300 - 1000 = 300$$

units of quantity. The new transportation problem is

	W_1	W_2	W_3	Fictious	Available
F_1	(30)	(20)	(10)	(0)	800
F_2	(5)	(15)	(25)	(0)	500
Required	300	300	400	300	1300

Solving this by usual method, we have an optimal solution given by the following matrix

	W_1	W_2	W_3	Fictious	Available
F_1	(30)	100 (20)	400 (10)	300 (0)	800
F_2	300 (5)	200 (15)	(25)	(0)	500
Requirement	300	300	400	300	

T. C $=100 \times 20 + 400 \times 10 + 300 \times 0 + 300 \times 5 + 200 \times 15 = 10{,}500$

This matrix show that 300 units will be left at F_1 not dispatched.

Example 9. solve the transportation problem

Warehouse → Factory ↓	W_1	W_2	W_3	W_4	Capacity
F_1	(19)	(30)	(50)	(10)	7
F_2	(70)	(30)	(40)	(60)	9
F_3	(40)	(8)	(70)	(20)	18
Warehouse Requirement	5	8	7	14	

[R. U. M. Sc. Math. 81, Shivaji M. Sc. Math. 76, Meertu M. Com. 72, Delhi 71]

Solution : We solve this problem by the Vogel's approximation method.

	W_1	W_2	W_3	W_4	Available	Penalties
F_1	(19)	(30)	(50)	(10)	7	(9)
F_2	(70)	2 (30)	(40)	(60)	9	(10)
F_3	(40)	6 (8)	(70)	(20)	10	(12)
Requirement	5	8	7	14		
Penalties	(21)	(22) ↑	(10)	(10)		

Maximum penalty is associated with column 2. Allocate 6 to (3,2) and 2 to (2, 2). Then the requirement of column 2 is satisfied. So omit this column and get the following reduced matrix

	W_1	W_3	W_4	Available	Penalties
F_1	5 (19)	(50)	(10)	7	(9)
F_2	(70)	(40)	(60)	7	(20)
F_3	(40)	(70)	(20)	12	(20)
Requirement	5	7	14		
Penalties	(21) ↑	(10)	(10)		

Allocate 5 to (1, 1). Then requirement of column first is satisfied.

So leave the column 1 and set the new reduced matrix as follow

	W_3	W_4	Available	Penalties
F_1	(50)	(10)	2	(40)
F_2	(40)	(60)	7	(20)
F_3	(70)	12 (20)	12	← (50)
Requirement	7	14		
Penalties	(10)	(10)		

Allocate 12 to (3, 2) of this reduced matrix. Capacity of F_3 is exhausted, omit F_3 and find the following reduced matrix.

	W_3	W_4	Available	Penalties
F_1	(50)	2 (10)	2	40
F_2	7 (40)	(60)	7	20
Requirement	7	2		
Penalties	100	50		
		↑		

The full allocations are shown in the following table.

	W_1	W_2	W_3	W_4	Available
F_1	5 (19)	(30)	(50)	2 (10)	7
F_2	(70)	2 (30)	7 (40)	(60)	9
F_3	(40)	6 (8)	(70)	12 (20)	18
Requirements	5	8	7	14	

The total transportation cost to this feasible solution is

$Z = 5 \times 19 + 2 \times 10 + 2 \times 30 + 7 \times 40 + 6 \times 8 + 12 \times 20 = 743$

Example. 10. Solve the transportation problem

	D_1	D_2	D_3	D_4	D_5	D_6	Available
O_1	(9)	(12)	(9)	(6)	(10)	(10)	5
O_2	(7)	(3)	(7)	(7)	(5)	(5)	6
O_3	(6)	(5)	(9)	(11)	(3)	(11)	2
O_4	(6)	(8)	(11)	(2)	(2)	(10)	9
b_j	4	4	6	2	4	2	

[Kanpur 71, Delhi B. Sc. Math. 75, M.Com. 75, Kurukshetra 76]

Solution : We use here the method of Lowest Cost Entry method. The Lowest minimum cost 2 is at (4, 4) and (4, 5).

	D_1	D_2	D_3	D_4	D_5	D_6	Available
O_1	(9)	(12)	5 (9)	(6)	(9)	(10)	5
O_2	(7)	4 (3)	(7)	(7)	(5)	2 (5)	6
O_3	1 (6)	(5)	1 (9)	(11)	(3)	(11)	2
O_4	3 (6)	(8)	(11)	2 (2)	4 (2)	(10)	9
	4	4	6	2	4	2	

Allocate maximum possible amount 2 and 4 respectively to these cells. Due to this allocations the columns 4 and 5 got required demands. Next lowest cost 3 is at (2, 2). Allocate maximum possible amount 4 to (2, 2). The requirement of column 2 is satisfied. Continuing in this way we find the allocations are as follows; 2 to (2, 6), 1 to (3, 1), 3 to (4, 1) 5 to (1, 3) and 1 to (3, 3).

Total transportation cost is

$= 5 \times 9 + 4 \times 3 + 2 \times 5 + 1 \times 6 + 1 \times 9 + 3 \times 5 + 2 \times 2 + 4 \times 2$

$= 112$

Example 11. Solve the transportation problem.

Markets → Plants ↓	A	B	C	D	Available
X	(10)	(22)	(10)	(20)	8
Y	(15)	(20)	(22)	(8)	13
Z	(20)	(12)	(10)	(15)	11
Requirement	5	11	8	8	

[Meerut M. Com. 74]

Solve : Adopting the vogel's approximation method, we get

	A	B	C	D	Available	Penalties
X	5 (10)	(22)	3 (10)	(20)	8	←(10)
Y	(15)	(20)	(12)	(8)	13	(4)
Z	(20)	(12)	(10)	(15)	11	(2)
Requirements	5	11	8	8		
Penalties	(5)	(8)	(2)	(7)		

Striking of the first column we get the reduced matrix.

	B	C	D	Available	Penaltie
Y	(20)	5(12)	8(8)	13	4
Z	11(12)	(10)	(15)	11	2
Requirements	11	5	8		
Penalties	8	2	7		

The Final allocations are shown in the following matrix.

	A	B	C	D	Available
X	5 (10)	(22)	3 (10)	(20)	8
Y	(15)	(20)	5 (12)	8(8)	13
Z	(20)	11(12)	(10)	(15)	11
Requirements	5	11	8	8	

The total transportation cost is

$$= 5 \times 10 + 3 \times 10 + 5 \times 12 + 8 \times 8 + 11 \times 12 = 336$$

Example 12. Solve the following transportation problem—

To

From		A	B	C	Available
	I	(2)	(7)	(4)	5
	II	(3)	(3)	(7)	8
	III	(5)	(4)	(1)	7
	IV	(1)	(6)	(2)	14
Requirement		7	9	18	

[R. U., B. Sc. Special Hons. 88 H, 90 H]

Solution :

To

From		A	B	C	Available	Penalty
	I	(2)	(7)	(4)	5	(4 – 2 = 2)
	II	(3)	8 (3)	(7)	8	←(7 – 3 = 4)
	III	(5)	(4)	(1)	7	(4 – 1 = 3)
	IV	(1)	(6)	(2)	14	(2 – 1 = 1)
Requirement : →		7	9	18		
Penalty : →		(2 – 1 = 1)	(4 – 3 = 1)	(2 – 1 = 1)		

Strike off the row II, then the reduced matrix is

To

	A	B	C	Available	Penalty
I	(2)	(7)	(4)	5	(4 – 2 = 2)
III	(5)	(4)	7 (1)	7	(4 – 1 = 3)
IV	(1)	(6)	(2)	14	(2 – 1 = 1)
Requirement	7	1	18		
Penalty	(2 – 1 = 1)	(6 – 4 = 2)	(2 – 1 = 1)		

Strike off the row III. The reduced matrix is shown below.

To

From		A	B	C	Available	Penalty
	I	(2)	(7)	(4)	5	(4 – 2 = 2)
	IV	(1)	(6)	11 (2)	14	(2 – 1 = 1)
Requirement :—		7	1	11		
Penalty		(2 – 1 = 1)	(7 – 6 = 1)	(4 – 2 = 2)		

Now, strike off the column 3, the reduced matrix is

	To			
	A	*B*	Available	Penalty
I	5 (2)	(7)	5	(7 – 2 = 5)
IV	2 (1)	1 (6)	3	← (6 – 1 = 5)
Requirement	7	1		
Penalty	(2 – 1 = 1)	(7 – 6 = 1)		

The full allocation is given by

		To			
		A	*B*	*C*	Available
	I	5 (2)	(7)	(4)	5
From	*II*	(3)	8 (3)	(7)	8
	III	(5)	(4)	7 (1)	7
	IV	2 (1)	1 (6)	11(2)	14
Requirement		7	9	18	

$$\text{T.C.} = 5 \times 2 + 8 \times 3 + 7 \times 1 + 2 \times 1 \times 6 + 11 \times 2$$
$$= 10 + 24 + 7 + 2 + 6 + 22$$
$$\text{T.C.} = 71$$

Example 13. Solve the following problem—

		To					
		A	*B*	*C*	*D*	*E*	Available
	I	(3)	(4)	(6)	(8)	(9)	20
From	*II*	(2)	(10)	(1)	(5)	(8)	30
	III	(7)	(11)	(20)	(40)	(3)	15
	IV	(2)	(1)	(9)	(14)	(16)	13
Requirement		40	6	8	18	6	

[Delhi M. Sc. Math. 68, Meerut 75]

Solution—

		To						
		A	*B*	*C*	*D*	*E*	Available	Penalty
	I	(3)	(4)	(6)	(8)	(9)	20	(4 – 3= 1)
From	*II*	(2)	(10)	8 (1)	(5)	(8)	30	(2 – 1 = 1)
	III	(7)	(11)	(20)	(40)	(3)	15	(7 – 3 = 4)
	IV	(2)	(1)	(9)	(14)	(16)	13	(3 – 1 = 1)
Requirement		40	6	8 ↑	18	6		
Penalty		(3 – 2 = 1)	(4 – 1 = 3)	(6 – 1 = 5)	(8 – 5 = 3)	(8 – 3 = 5)		

Now Striking off the column C, we get

To

		A	B	D	E	Available	Penalties
	I	(3)	(4)	(8)	(9)	20	(4 – 3 = 1)
From	II	(2)	(10)	(5)	(8)	22	(5 – 2 = 3)
	III	(7)	(11)	(40)	6 (3)	15	(7 – 3 = 4)
	IV	(2)	(1)	(14)	(16)	13	(2 – 1 = 1)
Requirements		40	6	18	6 ↑		
Penalties		(3 – 2 = 1)	(4 – 1 = 3)	(8 – 5 = 3)	(8 – 3 = 5)		

Now, strike off the column E. The reduced matrix is

To

		A	B	D	Available	Penalties
	I	(3)	(4)	(8)	20	(4 – 3 = 1)
From	II	(2)	(10)	(5)	22	(5 – 2 = 3)
	III	9 (7)	(11)	(40)	15	←(11 – 7 = 4)
	IV	(2)	(1)	(14)	13	(2 – 1 = 1)
Requirements		40	6	18		
Penalties		(3 – 2 = 1)	(4 – 1 = 3)	(8 – 5 = 3)		

Now, strike off the III row. The reduced matrix is

To

		A	B	D	Available	Penalties
	I	(3)	(4)	(8)	20	(4 – 3 = 1)
From	II	(2)	(10)	(5)	22	(5 – 2 = 3)
	III	(2)	6 (1)	(14)	13	(2 – 1 = 1)
Requirement		31	6	18		
		(3 – 2 = 1)	(4 – 1 = 3)	(8 – 5 = 3)		

Now, Strike off the column B. We have

To

		A	D	Available	Penalties
	I	(3)	(8)	20	(8 – 3 = 5)
From	II	(2)	(5)	22	(5 – 2 = 3)
	IV	7 (2)	(14)	7	(14 – 2 = 12)
Requirement		31	18		
Penalties		(3 – 2 = 1)	(8 – 5 = 3)		

Now, strike off the IV row.

To

		A	D	Available	Penalties
	I	20 (3)	(8)	20	(8 – 3 = 5)
From	II	4 (2)	18 (5)	22	(5 – 2 = 3)
Requirement		24	18		
Penalties		(3 – 2 = 1)	(8 – 5 = 3)		

Now, the full allocation is given by

To

		A	B	C	D	E	Available
	I	20 (3)	(4)	(6)	(8)	(9)	20
From	II	4 (2)	(10)	8 (1)	18 (5)	(8)	30
	III	9 (7)	(11)	(20)	(40)	6 (3)	15
	IV	7 (2)	6 (1)	(9)	(14)	(16)	13
Requirement		40	6	1	18	6	

$$\text{Optimal T. C.} = 20 \times 3 + 4 \times 2 + 8 \times 1 + 18 \times 5 + 9 \times 7 + 6 \times 3$$
$$+ 7 \times 2 + 6 \times 1 = 267$$

Optimal T. C.= 331

Example 14. Solve the following transportation problem.

	X	Y	Z	Supply
A	(8)	(7)	(3)	60
B	(3)	(8)	(9)	70
C	(11)	(3)	(5)	80
Demand	50	80	80	

Solution : Using Vogel's approximation method.

	X	Y	Z	Supply	PeInalty
A	(8)	(7)	(3)	60	4
B	50 (3)	(8)	(9)	70	←5
C	(11)	(3)	(5)	80	2
Demand	50	80	80		
Penalty	5	4	2		

Strike of the first column then the reduced matrix is

	Y	Z	Supply	Penalty
A	(7)	60 (3)	60	←4
B	(8)	(9)	20	1
C	(3)	(5)	80	2
Demand	80	80		
Penalty	4	2		

Strike off the first row. Then the reduced matrix is

	Y	Z	Supply	Penalty
B	(8)	20 (9)	20	1
C	80 (3)	(5)	80	2
Demand	80	20		
	↑			
Penalty	5	4		

The full allocation is

	X	Y	Z	Supply
A	(8)	(7)	60 (3)	60
B	50 (3)	(8)	20 (9)	70
C	(11)	80 (3)	(5)	80
Demand	50	80	80	

T. C. = $60 \times 3 + 50 \times 3 + 20 \times 9 + 80 \times 3 = 750$

Example 15. A company has four points P_1, P_2, P_3, P_4 from which is supplies to three markets M_1, M_2, M_3 Determine the transportations cost from the following data giving the plant to market shifting costs, quantities available as each plant and quantity required to each market.

Market	Plant				Required at market
↓	P_1	P_2	P_3	P_4	
M_1	(19)	(14)	(23)	(11)	11
M_2	(15)	(16)	(12)	(21)	13
M_3	(30)	(25)	(16)	(39)	19
Available at Plant	6	10	12	15	43

[R..U. Condensed Course Hons. 89]

Solution : Using Vogel's approximation method.

	P_1	P_2	P_3	P_4	Available	Penalty
M_1	(19)	(14)	(23)	11(11)	11	3
M_2	(15)	(16)	(12)	(21)	13	3
M_3	(30)	(25)	(16)	(39)	19	9
b_j	6	10	12	15		
Penalty	4	2	4	10 ↑		

Strike of first row. Then the reduced matrix is

	P_1	P_2	P_3	P_4	Available	Penalty
M_2	(15)	(16)	(12)	4 (21)	13	3
M_3	(30)	(25)	(16)	(39)	19	9
b_j	6	10	12	4		
Penalty	15	9	4	18 ↑		

Strike off 4th column. The reduced matrix is

	P_1	P_2	P_3	Available	Penalty
M_2	6 (15)	(16)	(12)	9	3
M_3	(30)	(25)	(16)	19	9
bj	6	10	12		
Penalty	15 ↑	9	4		

Strike off first column. The reduced matrix is

	P_2	P_3	Available	Penalty
M_2	3 (16)	(12)	3	4
M_3	7 (25)	12 (16)	19	9
b_j	10	12		
Penalty	9 ↑	4		

The full allocation is

	P_1	P_2	P_3	P_4	Available
M_1	(19)	(14)	(23)	11(11)	11
M_2	6 (15)	3 (16)	(12)	4(21)	1
M_3	7 (30)	7 (25)	12 (16)	(39)	19
b_j	6	10	12	15	

T. C. = 11 × 11 + 6 × 15 + 3 × 16 + 4 × 21 + 7 × 25 + 12 × 16
= 710

Example 16. Solve the following transportation problem

	A	B	C	D	Supply
X	(2)	(16)	(25)	(13)	11
Y	(17)	(18)	(14)	(3)	13
Z	(32)	(27)	(18)	(41)	19
Demand	6	10	12	15	

[Delhi M. Com. 76, 90, Meerut 76, 79 (S), M. Sc. (Stat) 75]

Solution : Using Vogel's approximation method

	A	B	C	D	Supply	Penalty
X	(21)	(16)	(25)	11 (13)	11	3
Y	(17)	(18)	(14)	(23)	13	3
Z	(32)	(27)	(18)	(41)	19	9
Demand	6	10	12	15		
Penalty	4	2	4	10 ↑		

Strike off irst row. The reduced matrix is

	A	B	C	D	Supply	Penalty
Y	(17)	(18)	(14)	4 (23)	13	3
Z	(32)	(27)	(18)	(41)	19	9
Demand	6	10	12	4		
Penalty	15	9	4	18 ↑		

Now, the reduced matrix is

	A	B	C	Supply	Penalty
Y	(17)	(18)	(14)	← 9	3
Z	(32)	(27)	12 (18)	← 19	9
Demand	6	10	12		
Penalty	15	9	4		

The next reduced matrix is

	A	B	Supply	Penalty
Y	6 (17)	3 (18)	9	1
Z	(32)	7 (27)	7	5
Demand	6	10		
Penalty	15 ↑	9		

The full allocation is

	A	B	C	D	Supply
X	(21)	(16)	(25)	11 (13)	11
Y	6 (17)	3 (18)	(14)	4 (23)	13
Z	(32)	7 (27)	12 (18)	(41)	19
Demand	6	10	12	8	

$$\text{T.C.} = 11 \times 13 + 6 \times 17 + 3 \times 18 + 4 \times 23 + 7 \times 27 + 12 \times 8 = 796$$

Optimality Test

After getting the initial basic feasible solution of a transportation problem, we test this solution for optimality i.e. we check whether the solution thus obtained minimizes the total transportation cost or not. The optimality test is applicable to a feasible solution consisting of (m + n – 1) allocations in independent positions.

Computational Pocedure of Optimality Test

Find the initial B.F.S. of a transportation problem and test for optimality as follows:

(i) For a B.F.S., we determine a set of (m + n) numbers

ui = 1, 2, ..., m

vj, = j = 1, 2, ..., n

such that, for each occupied cells (r, s), $c_{rs} = u_r + v_s$

For this we assign an arbitrary value to one of the u_i's or v_j's then the rest (m + n – 1) of them can easily be solved from the relation $c_{rs} = u_r + v_s$, for occupied cells. Generally we choose that u_i or $v_j = 0$ for which the corresponding row or column have the maximum number of individual allocations.

(ii) Then we calculate the cell evaluations d_{ij} for each unoccupied cell (i, j) by using the formula $d_{ij} = c_{ij} - (u_i + v_j)$

(iii) After getting dij following conclusions are made

(a) If all $d_{ij} > 0$ then the solution under test is optimal and unique.

(b) If all $d_{ij} \geq 0$, with at least one $d_{ij} = 0$, then the solution under test is optimal and alternative solution exists.

(c) If at least one $d_{ij} < 0$, then the solution is not optimal. In the last case we proceed to the next step (iv).

(iv) If at least one $d_{ij} < 0$, in step (iii), then we form new B.F.S. In the next B.F.S. we give maximum allocation to the cell for which d_{ij} is minimum and negative, by making an occupied cell empty.

Then we repeat the steps (i) to (iii) to test the optimality to this new B.F.S.

MODI Method for Transportations Algorithm

Step 1: Construct a transportation table entering the capacities $a_1, a_2, \dots a_m$ of the sources and the requirements $b_1, b_2 \dots b_n$. Enter the various costs c_{ij} at the upper left corners of all the cells. Find an initial B.F. solution allocation in independent positions) of the problem by Vogel method or by any methods. Enter the allocations at the centers of the cells.

Step 2: Find the set of numbers u_i $(i = 1, 2, \dots m)$ and $v_j (j = 1, 2, \dots n)$ by the method given in step 2 such that for each occupied cell (r, s), $c_{rs} = u_r + v_s$

Step 3: Find the cell evaluation $u_i + v_j$ for each unoccupied cell (i,j) and enter at the upper right corner of the corresponding cell (i, j).

Step 4: Find the cell evaluations $d_{ij} = c_{ij} - (u_i + v_j)$ for each unoccupied cell (i, j) and enter at the lower right corner of the corresponding cells.

Step 5: Examine the cell evaluations d_{ij} for unoccupied cells and conclude that

(i) If all $d_{ij} > 0$, then the solution under test is optimal and unique.

(ii) If all $d_{ij} \geq 0$, with at least one $d_{ij} = 0$, then the solution under test is optimal and alternative optimal solution exists.

(iii) If at least one $d_{ij} < 0$, then the solution is not optimal. In the last case we proceed to the next step 6.

Step 6: From a new B.F.S by giving maximum allocation to the cell for which d_{ij} is minimum and negative, by making an occupied cell empty.

Step 7: Then repeat the step 2 to 5, to test the optimality to this new B.F. solution. Continue improving the B.F.S iteratively using the step 2 to 6 till an optimal solution is attained.

Thus in the table after making all the entries the occupied (cells having allocations) and the unoccupied cells will be a follows.

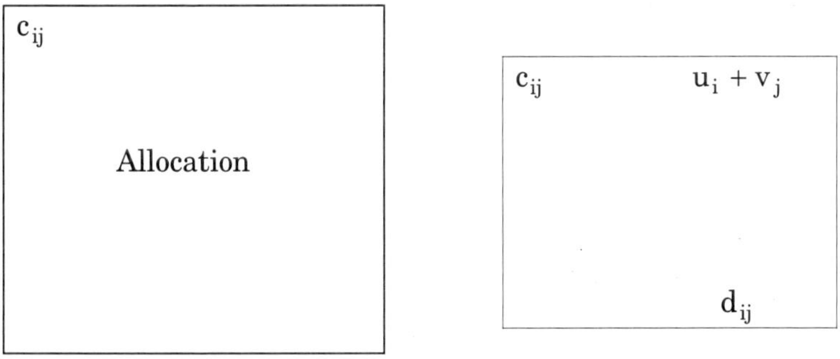

Example for Optimality Test

Example 1: A company has four plants P_1, P_2, P_3, P_4 from which it supplies to three markets M_1, M_2, M_3. Determine the optimal transportation plan from the following data giving the plant to market shifting costs, quantities available at each plant and quantities required at each market.

Market	Plant				Required at Market
	P_1	P_2	P_3	P_4	
M_1	19	14	23	11	11
M_2	15	16	12	21	13
M_3	30	25	16	39	19
Available at Plant Plant	6	10	12	15	43

Solution:

Step 1: By V.A.M. Method an initial B.F.S of the given problem is given by the following table

Table I

19	14	23	**11** 11	11
6 15	**3** 16	12	**4** 21	13
30	**7** 25	**12** 16	39	19
6	10	12	15	

Total transportation cost = 6*15 + 3*16 + 7*25 + 12*16 + 11*11 + 4*21 =Rs 710

Step 2: Here we determine a set of u_i and v_j s.t. for each occupied cell (r, s), $c_{rs} = u_r + v_s$.

For this, we choose $u_2 = 0$ (since row 2 contains maximum number of allocations) since

$$u_{21} + 15 = u_2 + v_1$$

$$u_{22} = 16 = u_2 + v_2$$

$$u_{24} = 21 = u_2 + v_4$$

$$\therefore\ v_1 = 15, v_2 = 16, v_4 = 21$$

Also $u_{14} = 11 = u_1 + v_4$

$$u_{32} = 25 = u_3 + v_2$$

$$u_{33} = 16 = u_3 + v_3$$

$$u_1 = 11 - v_4 = -10,$$

$$u_3 = 25 - v_2 = 9$$

$$v_3 = 16 - u_3 = 7$$

Step 3: Now we find the cell evaluations $u_i + v_j$ for each occupied cell (i, j) and enter at the upper right corner of the corresponding unoccupied cell.

Step 4: Then we find the cell evaluations $d_{ij} = c_{ij} - (u_i + v_j)$ (i.e. the difference of upper right corner entry from the upper left corner entry) for each occupied cell (i, j) and enter at the lower right corner of the corresponding unoccupied cell (i, j).

Thus we get the following table

Table II

					ui
	19 5 14	14 6 8	23 -3 26	11 **11**	- 10(=u1)
	15 **6**	16 **3**		21 **4**	0(=u2)
	30 24 6	25 7	16 **12**	39 30 9	9(=u3)
	15	16	7	21	
vj	v_1	v_2	v_3	v_4	

Step 5: Since all $d_{ij} > 0$, hence the B.F.S. shown by table II is an optimal solution and the optimal transportations cost = 710 Rs. Thus the solution of the given transportations problem is

From source 1 transport 11 units to destination 4
From source 2 transport 6, 3 and 4 units to destination 1, 2, and 4 respectively.
And From source 3 transport 7, and 12 units to destinations 2 and 3 respectively.

Example 2: obtain optimal Transportation cost for the following Transportation problem.

		To		Supply
From	1	2	3	
1	2	7	4	5
2	3	3	1	8
3	5	4	7	7
4	1	6	2	14
Demand	7	9	18	34

Step 1: The initial B.F.S. of the above problem by Vogel Method
Total Transportation cost = 5*2 + 2*1 + 7*4 + 2*6 + 8*1 + 10*2 = Rs. 80

Table 1

				a_i
	2 **5**	7	4	5
	3	3	1 **8**	8
	5	4 **7**	7	7
	1 **2**	6 **2**	2 **10**	14
b_j	7	9	18	

Step 2: Now we determine a set of u_i and v_j s.t for each occupied cell (r, s), $c_{rs} = u_r + v_s$. For this, we choose $u_4 = 0$ (since row 4 contains maximum number of allocations)

Since $c_{41} = 1 = u_4 + v_1$

$c_{42} = 6 = u_4 + v_2,$

$c_{43} = u_4 + v_3 = 2$

$\therefore$ $v_1 = c_{41} - u_4 = 1,\ v_2 = 6 - u_4 = 6,\ v_3 = 2 - u_4 = 2$

Also $c_{11} = 2 = u_1 + v_1,\ c_{23} = 1 = u_2 + v_3,\ c_{32} = 4 = u_3 + v_2$

$u_1 = 2 - v_1 = 1,\ u_2 = 1 - v_3 = -1,\ u_3 = 4 - v_2 = -2$

Step 3: Then we find the cell evaluations $(u_i + v_j)$ for each unoccupied cell (i, j) and enter at the upper right corner of the corresponding unoccupied cell.

Step 4: Then we find the cell evaluations $d_{ij} = c_{ij} - (u_i + v_j)$ (i.e. the difference of upper right corner entry from the upper left corner entry) for each unoccupied cell (i, j) and enter at the lower right corner of the corresponding unoccupied cell.

Thus we get the following table:

Table-2

				u_i
	2 **5**	7 7 **0**	4 3 1	1(=u_1)
	3 0 **3**	3 5 –2	1 **8**	–1(=u_2)
	5 -1 6	4 7	7 0 7	-2(=u_3)
	 2	6 2	2 **10**	0(=u_4)
v_j	$v_1 = 1$	$v_2 = 6$	$v_3 = 2$	

Step 5: Since cell evaluation $d_{22} = -2 < 0$, so the solution under test is not optimal.

Step 6: Since minimum d_{ij} is $d_{22} = -2$ (negative), so we give maximum allocation to this cell from an unoccupied cell and make the necessary changes in other allocations as shown in table 3.

Table 3

5		
	+2	–2
2	–2	+2 10

Step 7: The new B.F.S. (allocations in independent positions) thus obtained is shown in Table 4. For this B.F.S total transportation cost = 5*2 + 2*1 + 2*3 + 7*4 + 6*1 + 12*2 = Rs 76, which is less than for the initial B.F.S.

Table 4

				ai
	2 **5**	7	4	5
	5	3 **2**	1 **6**	8
	5	4 7	7	7
	1 **2**	6	2 **12**	14
b_j	7	9	18	

Step 8: Proceeding as in step 2, 3, and 4 we get the following table:

Table 5

			u_i
2 5	7 5 2	4 3 1	1(=u_1)
3 0 3	3 2	1 6	–1(=u_2)
5 1 4	4 7	7 2 5	0(=u_3)
1 2	6 4 2	2 12	0(=u_4)
1	4	2	

Since all $d_{ij} > 0$, hence the B.F.S shown by table 5, an optimal solution which is also unique.

Thus the solution of the given transportation problem is

From source 1 transport 5 units to destination 1

From source 2 transport 2, and 6 units to destinations 2, and 3 respectively.

From source 3 transport 7 units to destination 2.

and from source 4, transport 2, and 12 units to destinations 1 and 3 respectively and the total transportations cost (optimal) = Rs. 76

Example 3: Find the optimal transportations cost for the following transportation problem.

	S_1	S_2	S_3	S_4	a_i
O_1	1	2	1	4	30
O_2	3	3	2	1	50
O_3	4	2	5	9	20
b_j	20	40	30	10	100

Solution:

By lowest cost method, we get the following B.F.S. of the problem.

Table 1

	S_1	S_2	S_3	S_4	a_i
O_1	1 **20**	2	1 10	4	30
O_2	3	3 **20**	2 **20**	1 **10**	50
O_3	4	2 **20**	5	9	20
b_j	20	40	30	10	100

and the total transportations cost = 20*1 + 10*1 + 20*3 + 20*2 + 10*1 + 20*2 = Rs 180

To Test the Solution For Optimality:

Finding the set of $u_i(i = 1, 2, 3)$, $v_j = (j = 1, 2, 3, 4)$, such that $c_{rs} = u_r + v_s$ for occupied cells and entering $u_i + v_j$ and d_{ij} in the unoccupied cells, the table giving the optimal solution is as follows. (see table 2)

Since all $d_{ij} \geq 0$, the solution given by the above table is an optimal solution. Since $d_{12} = 0$, it implies that solution is not unique but an alternative optimal solution will exist. Optimal transportations cost Rs 180.

Table 2

					u_i
	1 **20**	2 2 0	1 **10**	4 0 4	$-1(= u_1)$
	3 2 1	3 **20**	2 20	1 **10**	$0(= u_2)$
	4 1 3	2 **20**	5 1 4	9 0 9	$-1(= u_3)$
v_j	$v_1 = 2$	$v_2 = 3$	$v_3 = 2$	$v_4 = 1$	

Thus the solution of the problem is

From source O_1 transport 20 and 10 units to destination S_1 and S_2 respectively.

From source O_2 transport 20, 20 and 10 units to destinations S_2, S_3 and S_4 respectively.

From source O_3 transport 20 units to destination S_2.
And the total transportation cost = Rs 180.

EXERCISE

1. Describe a method to obtain an initial feasible solution for a transportation problem. [Meerut Math. 72]
2. Explain North-West Corner Rule. [Kanpur 70]
3. Give an algorithm for solving transportation problem.
4. Write a short note on "Transportation problem". [Meerut O. R. 73, 74, 79]
5. Explain the difference between a transportation problem and as assignment problem. [Meerut O. R. 70, 73 (S), 75]
6. Give a brief outline of the procedure for solving a transportation problem. [Meerut O. R. 77]
7. Solve the following transportation problem.

(i)

		To 1	2	3	Available
From	*I*	(2)	(7)	(4)	5
	II	(3)	(3)	(7)	8
	III	(5)	(4)	(1)	7
	IV	(1)	(6)	(2)	14
Requirement		7	9	18	

[Delhi 70]

Ans. *I* to 1 units 5
II to 2 units 8
III to 3, 7 units
IV to 1 units 2, to 2 units 1,
T.C. = 71 to 3 units 11.

(ii)

Ware house → Plants ↓	↓	*A*	*B*	*C*	*D*	Available
	X	(30)	(20)	(50)	(20)	75
	Y	(20)	(10)	(20)	(40)	120
	Z	(40)	(30)	(40)	(30)	105
Requirement		65	60	80	95	

[Meerut M. Com. 75]

Ans.	X	to D units 75
	Y	to B units 40
		to C units 80
	Z	to A units 65
		to B units 70
		to D units 20

Total Transport cost = 7300

(iii) To

		W_1	W_2	W_3	W_4	W_5	Available
	F_1	(3)	(4)	(6)	(8)	(9)	20
From	F_2	(2)	(10)	(1)	(5)	(8)	30
	F_3	(7)	(11)	(20)	(40)	(3)	15
	F_4	(2)	(1)	(9)	(14)	(10)	13
		40	6	8	18	6	

[Meerut 75]

Ans.	F_1	to W_1 units 20
	F_2	to W_1 units 4
		to W_3 units 8
		to W_4 units 18
	F_3	to W_1 units 9
		to W_5 units 6
	F_4	to W_1 units 7
		to W_2 units 6 T.C. = 267

(iv) To

						Available
	(3)	(4)	(6)	(8)	(9)	20
	(2)	(10)	(1)	(5)	(8)	30
From	(7)	(11)	(20)	(40)	(3)	15
	(2)	(1)	(9)	(14)	(16)	13
Requirement	40	6	8	18	6	

[Delhi M. Sc. Math. 68, Meerut 75]

Ans.

20				
4		8	18	
9				6
7	6			

T. C. = 267 units

(V) Market

		A	B	C	Available
	X	(11)	(21)	(16)	14
Plant	Y	(7)	(17)	(13)	26
	Z	(11)	(23)	(21)	36
Required at Market		18	28	25	

[Meerut M. Com. 75]

Ans. *X* to *C* units 14

Y to *B* units 15

to *C* units 18

Z to *A* units 18

to *B* units 13

5 units are left at *Z*

Total T. Cost = 119

8. A manufacturer wants to ship 8 loads of his product as shown below. The matrix gives the mileage from origin *O* to the destinations.

D → ↓ *O*	A	B	C	Available
X	(50)	(30)	(220)	1
Y	(90)	(45)	(170)	3
Z	(250)	(200)	(50)	4
Requirement	4	2	2	

Shipping cost are Rs. 10 per load mile. What shipping schedule should be used?

[Punjab M. Sc. Math. 75, Delhi 71, Kanpur 69]

Ans. *X* to *A* units 1, *Y* to *A* units 3, *Z* to *B* units 2 and to *C* units 2, costs Rs. 8200.

Solve the following transportation problem.

D → ↓ *O*	A	B	C	Available
1	(2)	(7)	(4)	5
2	(3)	(3)	(1)	8
3	(5)	(4)	(7)	7
4	(1)	(6)	(2)	14
Requirement b_j	7	9	18	

[Meerut 70. M. Com. 77, Dibrugarh M. Sc. Stat. 79]

Ans.

5		
		8
	7	
2	2	10

T. C. = 80

10. Solve the transportation problems.

	X	Y	Z	Available a_i
A	(8)	(7)	(3)	60
B	(3)	(8)	(9)	70
C	(11)	(3)	(5)	80
bj	50	80	80	

[C. A. Nov. 77, Delhi Math. 69, Trivendrum M. Sc. Elect, 77, Shivajee 77, Agra 84, Rohilkhand 83, 86, Delhi M. A. (Eco.) 1978]

Ans $\begin{bmatrix} 0 & 0 & 60 \\ 50 & 0 & 20 \\ 0 & 80 & 0 \end{bmatrix}$ T. C. = 750 units

11. Solve transportation problem

	Market A	B	C	D	Available
X	(14)	(9)	(18)	(6)	11
Y	(10)	(11)	(7)	(16)	13
Z	(25)	(20)	(11)	(34)	19
Required	6	10	12	15	

[Meerut M. Com. 73]

Ans. *X* to *D* units 11

Y to *B* units 6

to D units 4

Z to *B* units 7

to *C* units 12

Total T. Cost = Rs. 495.00

5

Assignment Problems

5.1. INTRODUCTION

If there are number of jobs (or tasks) which are to be assigned to an equal number of facilities (or people or machine) such that the assignment of a job to a particular person (or machine) would incur certain cost, then the problem of determining a set or sets of alternative assignment which would lead to an optimal cost is termed as an Assignment Problem. Here, we make the assumption that each person can perform each job but with varying degree of efficiency, for example, departmental head may have five persons available for assignment and five jobs to fill. Then his interest is to find the best assignment which will be in the best interest of the department.

5.2. ASSIGNMENT PROBLEM

Let three be m jobs for n number of persons. Let the assignment of the i-th job to j-th person would incur a cost c_{ij}. We assume that a particular job can only be assigned to particular person and conversely a particular person is capable of doing not more than one job at a time, we define the decision variable $x_{ij} \geq 0$ which would indicate the assignment of the i-th job to the j-th person. The objective is to minimize the total cost.

i.e. $$\text{Minimize total cost } z = \sum_{i=1}^{m} \sum_{j=1}^{n} c_{ij} x_{ij}$$

$$\text{Where } x_i = \begin{cases} \text{if the } i\text{–th job is assigned to the } j-\text{th person.} \\ 0 \text{ if the } i\text{–th job is not assigned to the } j\text{–th person.} \end{cases}$$

$i = 1, 2, \ldots\ldots m; j = 1. 2. \ldots\ldots n$

Since for a fixed j (i. e. α particular person) only one job can be assigned, we must have

$$\sum_{i=1}^{m} x_{ij} = 1, j = 1, 2, \ldots\ldots n$$

i.e. for a fixed j, only one $x_{ij} = 1$ for certain value of i, and rest of $(n - 1)$ x_{ij} s are zeros. Similarly since for fixed i, a particular job can only be assigned to one particular person, we must have

$$\sum_{j=1}^{n} x_{ij} = 1, i = 1, 2, \ldots\ldots m,$$

Structurally we can write Assignment problem as

$$\text{Min.} z = \sum_{i=1}^{m} \sum_{j=1}^{n} c_{ij}\, x_{ij}.$$

Subject to $\sum_{i=1}^{m} x_{ij} = 1$

$$\sum_{j=1}^{n} x_{ij} = 1$$

$x_{ij} \geq 0$ and satisfies.

$$x_{ij} = \begin{cases} 1 \text{ if the } i\text{th job is assigned to } j-\text{th person}, \\ 0, \text{ if the } i\text{th job is not assigned to the } j-\text{th person} \end{cases}$$

$i = 1, 2, \ldots\ldots m$

$j = 1. 2. \ldots\ldots. n$

(i) If $m = n$, then the problem is called *Balanced Assignment Problem.*

(ii) If $m \neq$ n, then the problem is known as *Unbalanced Assignment Problem.*

Note— The assignment problem can also be stated in the form of $n \times n$ matrix called the cost of Effectiveness matrix c_{ij} is the cost of assignment i-th facility to the j-th job. The effectiveness matrix $[C_{ij}]$ is

Job →		1	2	3	...	j	...	n
Person ↓	1	C_{11}	C_{12}	C_{13}	...	C_{1j}	...	C_{1n}
	2	C_{21}	C_{22}	C_{23}	...	C_{2j}	...	C_{2n}
	3	...	...	...	...	...	...	...
	4	...	...	...	...	...	...	...
$[C_{ij}] =$	:							
	:							
	i	C_{i1}	C_{i2}	C_{i3}	...	$C\text{-}_{ij}$	...	C_{in}
	:							
	n	C_{n1}	C_{n2}	C_{n3}	...	C_{nj}	...	C_{nn}

Theorem I. [R. U. 90 H]

Reduction Theorem

If in an assignment problem a constant is added or subtracted to every element of a row (or Column) of the cost matrix $[C_{ij}]$, then an assignment which minimizes the total cost for one matrix also minimizes the total cost for the other matrix. This may be put in the following. way—

If $x_{ij} = X_{ij}$ minimizes $Z = \sum_{i=1}^{n} \sum_{j=1}^{n} C_{ij} x_{ij}$

Over all x_{ij} such that

$$\sum_{i=1}^{n} x_{ij} = 1 = \sum_{j=1}^{n} x_{ij}$$

and $x_{ij} \geq 0$, then $x_{ij} = X_{ij}$ also minimizes

$$Z' = \sum_{i=1}^{n} \sum_{j=1}^{n} C'_{ij} X_{ij}$$

where $C'_{ij} = C_{ij} \pm di \pm kj$
where di, kj are constants; $i = 1, 2, \ldots\ldots\ldots\ldots, n$
$j = 1, 2, \ldots\ldots\ldots\ldots, n$

Proof : Here, we have

$$Z' = \sum_{i=1}^{n} \sum_{j=1}^{n} C'_{ij} X_{ij}$$

$$= \sum_{i=1}^{n} \sum_{j=1}^{n} (C_{ij} \pm d_i \pm k_j) x_{ij}$$

, putting the value of C'_{ij}

$$= \sum_{i=1}^{n} \sum_{j=1}^{n} c_{ij} x_{ij} \pm \sum_{i=1}^{n} \sum_{j=1}^{n} d_i x_{ij}$$

$$\pm \sum_{i=1}^{n} \sum_{j=1}^{n} k_j x_{ij}$$

$$= z \pm \sum_{i=1}^{n} di \sum_{j=1}^{n} x_{ij} \pm \sum_{i=1}^{n} k_j \sum_{j=1}^{n} x_{ij}$$

$$= Z \pm \sum_{i=1}^{n} d_i (1) \pm \sum_{j=1}^{n} k_j (1)$$

as $\sum_{i=1}^{n} x_{ij} = 1 = \sum_{j=1}^{n} x_{ij}$

$$= Z \pm \sum_{i=1}^{n} d_i \pm \sum_{j=1}^{n} k_j$$

Since $\sum_{i=1}^{n} d_i$ and $\sum_{j=1}^{n} k_j$ are independent of x_{ij}, therefore. Z' is minimizes when Z is minimized.

Hence, $x_{ij} = X_{ij}$ which minimizes Z, also minimizes Z'.

Theorem 2:

If all $C_{ij} \geq 0$ and there exists a solution $x_{ij} = X_{ij}$ such that

$$\sum_{i=1}^{n} \sum_{j=1}^{n} C_{ij} x_{ij} = 0.$$

Then this solution is an optimal solution (that is, this solution minimizes Z).

Proof: Given $C_{ij} \geq 0$

$$\therefore \qquad Z = \sum_{i=1}^{n} \sum_{j=1}^{n} C_{ij} \cdot x_{ij} \text{ can not be negative}$$

Thus its minimum value is 0, when $x_{ij} = X_{ij}$

Hence, the solution $x_{ij} = X_{ij}$

for which $\sum_{i=1}^{n} \sum_{j=1}^{n} C_{ij} X_{ij} = 0$ is an optimal solution.

Observation : (i) Keeping these theorems in mind we first create some zeroes in the given cost matrix and then try to find a complete assignment in terms of zeros. *By a Complete assignment we mean an assignment plan containing exactly n assigned independent zeroes, in one each row and one in each column.*

(ii) *Two zero are said to be independent if they lie in distinct rows and distinct columns.*

The Assignment problem (The Hungarian Method)

Case I : Problem of minimizing the total cost.

The Assignment algorithm is a step by step computational procedure for solving an assignment problem. This is a powerful method for the solution of the assignment problem which is in the from n x m matrix $[C_{ij}]$ called cost matrix. the procedure for the solution is as follows.

Step I :— Subtract the smallest element of each row in the cost matrix from that row. Then subtract the smallest element of each column from that column. These steps create zeros.

Step II :— Make assignments in terms of zeros. If a complete assignment is available, then this is the required optimal assignment, and if not, then we shall increase the number of zeros in the matrix. This can be down as follows.

(i) Starting with first row, examine rows of the matrix obtained in step 1 successively until a row containing exactly one zero in found, mark ❑ this zero, as an assignment will be made there.

(ii) Draw a vertical line though the column containing this assigned zero. This eliminates the confusion of making further assignments in that column. In this way examine all the rows.

(iii) After examining the rows completely, adopt a similar procedure to columns successively. Starting from first column, examine all the uncrossed columns until a column containing exactly one zero is found. Mark ❑ this zero and draw a horizontal line through the row containing this marked zero.

Adopting this method on rows and columns alternatively, we shall reach on an of the two situations.

(a)all the zeros have been marked (for decreased).

or (b)the remaining zeros be at least two in each row and column.

In the first case (I) we have assigned as much as we can write in the second case still we have some zeros to be treated. At this stage we generally use trial and error to break up such ties of zeros.

If the number of marked (assigned) zeros is exactly n, then we have obtained a complete optimal assignment. But if the number of assigned zeros is less than n, then we shall modify the cost matrix by creating some more zeros in it.

Remember that these lines thus drawn (horizontal and vertical both) are the minimum number of lines, required to pass through all the zeros of the matrix. It can be shown that the maximum number of independent zeros of the matrix is the same as the minimum number of lines required to pass through all the zero of the matrix. Thus if the number of the lines is less than n then the complete assignment is not possible, while if the number of lines is exactly n, then the complete assignment is obtained.

These vertical and horizontal lines drawn have the property that they cover all the zeros and every line pass through one and only one assignment (marked zero ❑) for, note that we have drawn lines through the columns and rows containing marked zero ❑, If a row contain this marked zero ❑, then it means we are assigning the job to two facilities which is a violation of the hypothesis. Hence, no line pass through more than one marked zero ❑

Step III :— If a complete assignment programme is not possible in step II, then select the smallest element out of those which do not lie on any of the lines in the above matrix. subtract this smallest element from al such uncrossed elements and add it to the elements which are placed at the intersection of horizontal and vertical lines. Don't alter the elements through which only one line passes. In this way a new matrix is formed and then apply step II to it.

If still a complete optimal assignment is not available in this new matrix, then use steps III and II iteratively.

Case II : Problem of maximizing the profit—

We can use the above procedure for problems of minimization by converting term into maximization by modifying the given cost matrix by either of the two methods.

(i) Select the greatest element of the given cost matrix and then subtract each element of the given cost matrix and then subtract each element of the matrix from this greatest element to get the modified matrix.

(ii) Multiply each element of the matrix by — 1 to get the modified matrix. Clearly it is covered to minimization problem.

Illustrative Example

Example 1. Solve the following minimal assignment problem

Man →		*I*	*II*	*III*	*IV*
Job	1	12	30	21	15
↓	2	18	33	9	31
	3	44	25	24	21
	4	23	30	28	14

[R. U. Condensed Course Hons. 89, Delhi 67]

To make the point clear we shall solve it in detail.

Here, we use the symbol (i, j) for the ith row and j-th column. For an example, zero at (2, 3) means zero at the intersection of 2nd row and 3rd column.

Step I : Subtract the smallest element of each row from every element of the corresponding row. The smallest element of the first row is 12. from other elements of first row. The elements of first row become

$$(12 - 12, 30 - 12, 21 - 12, 15 - 12) = (0, 18, 9, 3)$$

Similarly for the other row.

	I	*II*	*III*	*IV*
1	0	18	9	3
2	9	24	0	22
3	23	4	3	0
4	9	6	14	0

Now subtract the smallest element of each column of this table from every element of the corresponding column.

The smallest element of first column is 0. Subtract zero from every other elements of first column, that is 0—0, 9—0, 23—0, 9—0. Thus the elements of first column are 0, 9, 23, 9. Now the smallest element of 2nd column is 4, So the elements of 2nd column become 18—4, 24—4, 4—4, 16—4, that is 14, 20, 0, 12 similarly, we can find the elements of third and 4th column. Thus we get the matrix

	I	*II*	*III*	*IV*
1	0	14	9	3
2	9	20	0	22
3	23	0	3	0
4	9	12	14	0

Step II : Firstly we search out that row (or rows) which contained exactly one zero. We see that first row contain only one zero at (1, 1). We make assignment to this zero at (1, 1) by putting [0] and then draw a vertical line through (1, 1). The 2nd row has zero at (2, 3). Make assignment [0] to it and draw vertical line through it. 4th row contain zero at (4, 4). Make assignment [0] there and draw a vertical through it.

	I	*II*	*III*	*IV*
1	[0]	14	9	3
2	9	20	[0]	22
3	...23...	[0]	...3...	...0...
4	9	12	14	[0]

Examining all the rows. We now examine the columns. We see that uncrossed column is 2nd in which zero is at (3, 2). We assign

[0] these and draw a horizontal line through it. The number of lines drawn to cover all the zeros is equal to the order of the matrix which is 4. We see that every row and every column contain assignment ❑.

So we get the complete optimal zero assignment as follows.

Job	1	2	3	4
Man	I	II	III	IV

Example 2. Solve the following minimal assignment problem—

Man → *I* *I* *III* *IV*

Job ↓

Job	*I*	*I*	*III*	*IV*
1	2	3	4	5
2	4	5	6	7
3	7	8	9	8
4	3	5	8	4

[Meerut O. R.75 S]

Step I : Subtract the smallest element of each row from every element of the corresponding row. The smallest element of the first row is 2. subtract 2 from every element of the first rows. Then elements of first row become 2—2, 3—2, 4—2, 5—2.

i. e. 0, 1, 2, 3. Similarly for the other row. Thus we get the following matrix.

	I	*II*	*III*	*IV*
1	0	1	2	3
2	0	1	2	3
3	0	1	2	1
4	0	2	5	1

Now, Subtract the smallest element of each column from every element of the corresponding column, we get the following matrix.

	I	*II*	*III*	*IV*
1	0	0	0	2
2	0	0	0	2
3	0	0	0	0
4	0	1	3	0

Step II : We see that none of the rows or columns contain exactly one zero. so we start searching two zeros, we find that row 4 has zeros, We make assignment ❑ to this zero at (4, 1) and draw a vertical line through it. Note that no assignment can be made to the zero at (4, 4) because we have already made assignment in this row at (4.1). After drawing a vertical through (4, 1) we find that first row contains zeros at (1, 2) and at (1, 3). Make assignment ❑ to any one of them, say, at (1, 2) and draw a vertical through it. After drawing these two vertical lines, we find 2nd row contains zero at (2, 3). Make assignment [0] there and draw a vertical through it.

	I	II	III	IV
1	0	[0]	0	2
2	0	0	[0]	2
3	0	0	0	[0]
4	[0]	1	3	0

After drawing these three vertical lines we find that 3rd row contains zero at (3, 4). Make assignment ❑ there and draw vertical through it. Thus all the servos are covered by these four lines, giving assignment ❑ in every row and every column. Number of lines is equal to the order of matrix 4. Thus we get the following assignment.

$$1 — II, 2 — III, \qquad 3 — IV, 4 — I$$

Example 3. There are five jobs to be assigned, one each to 3 machines and the associated cost matrix is as follows—

Man → Job ↓	I	II	III	IV	V
A	11	17	8	16	20
B	9	7	12	6	15
C	13	16	15	12	16
D	21	25	17	28	26
E	14	10	12	11	15

[Indian Stat. Institute 71, M. Sc. Broada B. Sc. 78, Madras I. I.T. (M. Tech) 78, Delhi B. sc. Math 77, Sc. 71, M. B. A. 74]

Solution— Step I — Subtracting the smallest element of each row from its row, we get

	I	*II*	*III*	*IV*	*V*
A	3	9	0	8	12
B	3	1	6	0	9
C	1	4	3	0	4
D	4	7	0	11	9
E	4	0	2	1	5

Now, subtract the minimum element of each column from its column.

	I	*II*	*III*	*IV*	*V*
A	2	9	0	8	8
B	2	1	6	0	5
C	0	4	3	0	0
D	3	7	0	11	5
E	3	0	2	1	1

Step II — Now, we give the zero assignments in our usual manner and get the following matrix.

	I	*II*	*III*	*IV*	*V*
A	2	9	[0]	8	8
B	2	1	6	[0]	5
C	[0]	... 4 ...	... 3 ...	... 0 ...	... 0 ...
D	3	7	0	11	5
E	3	[0]	2	1	1

But row 4 and column 5 have got no assignment. so we proceed to next and get the following matrix which is obtained by the following method. The smallest of the elements that do not have a line through them is 1. Subtracting this element 1 from all the elements that do not have a line through them and adding to every element that lines at the intersection of two lines and leaving the remaining elements unchanged.

	I	II	III	IV	V
A	1	9	[0]	8	7
B	1	1	6	[0]	4
C	[0]	5	4	1	0
D	2	7	0	11	4
E	2	[0]	2	1	0

We see that row 4 and column V have again no assignment. So we repeat the above procedure and obtain the following matrix.

	I	II	III	IV	V
A	[0]	8	0	8	6
B	0	0	6	[0]	3
C	0	5	5	2	[0]
D	1	6	[0]	11	3
E	2	[0]	3	2	0

The optional assignment is

$$A \to I, B \to IV, C \to V, D \to III, E \to II$$

Example 4. A department head has four subordinates, and four tasks to be performed. The subordinate differ in efficiency and the tasks differ in the intrinsic difficulty. His estimate of the times each man would take to perform each task is given in the effectiveness matrix below. How should the task be allocated, one to a man so to minimize the total man hours?

Man →		I	II	III	IV
Job↓	A	8	26	17	11
	B	3	28	4	26
	C	38	19	18	15
	D	19	26	24	10

[Kurukshetra 75, Meerut Stat. 98, 70, Math 69, Delhi O. R. 64, Math 69, 70, B. A. Hons. 72, I. C., W. A, 79, Management accountancy Exami. (N.) Dec. 76. Sardar Patel Univ. Stat 75, Agra M. Stat 74 R. U. Part II Hons. 91 H, V. B. U. Part II Hons. 99]

Solution : Step I:— Subtracting the minimum element of each row from its row we get

	I	*II*	*III*	*IV*
A	0	18	9	3
B	9	24	0	22
C	23	4	3	0
D	9	16	14	0

Now, subtract the smallest element of each column from its column.

	I	*II*	*III*	*IV*
A	0	14	9	3
B	9	20	0	22
C	23	0	3	0
D	9	12	14	0

Step II — First row contains only one zero at (1, 1). Make assignment ❑ there and draw vertical through it. 2nd row contains zero at (2, 3). Make assignment ❑ there and draw vertical line through it. After examine rows, we now turn to uncrossed columns. We see that uncrossed column 2nd and 4th have zeros at (3, 2) and (4, 4) respectively. Put assignment ❑ there and draw horizontal lines through it.

Thus,

	I	*II*	*III*	*IV*
A	[0]	14	9	3
B	9	20	[0]	22
C	... 22 ...	[0]	... 3 ...	... 0 ...
D	... 9 ...	... 12 ...	... 14 ...	... [0]

by drawing 4 lines equal to the order of the matrix, all the zeros have been conversed; and every row and every column contains one assignment. So we have the complete optimal zero assignment as follows.

$$A \to I,\ B \to III,\ C \to II,\ D - IV$$

Example 5. Four men are available to do four different jobs. From past records, the time (in hours) that each man takes to do each job is know as given in the following table

Man → Job ↓	M_1	M_2	M_3	M_4
I	2	5	3	7
II	2	8	9	5
III	3	7	2	8
IV	1	5	3	1

Find the assignment of men to jobs that will minimize the total time taken.

Solution : Subtract the smallest element of each row from every element of the corresponding row. Then subtract the smallest element of each column from every element of the corresponding column.

	M_1	M_2	M_3	M_4
I	0	3	1	5
II	0	6	7	3
III	1	5	0	6
IV	0	4	2	0

	M_1	M_2	M_3	M_4
I	0	0	1	5
II	0	3	7	3
III	1	2	0	6
IV	0	1	2	0

Employing the procedure of assignments, we get the final assignment.

	M_1	M_2	M_3	M_4
I	… 0 …	$\boxed{0}$	… 1 …	… 5 …
II	$\boxed{0}$	3	7	3
III	… 1 …	… 2 …	$\boxed{0}$	… 6 …
IV	0	1	2	$\boxed{0}$

$$M_1 \to II, M_2 \to I, M_3 \to III, M_4 \to IV$$

The actual cost is clearly = 2 + 5 + 2 + 1 = 10 units and can be calculated by completing original matrix see the given table.

Assignment of M_1 is II which is at (2, 1) where cost is 2

"	" M_2 is *I*	"	"	(1, 2)	"	"	5
"	" M_3 is *III*	"	"	(3, 3)	"	"	2
"	" M_4 is *IV*	"	"	(4, 4)	"	"	1

Example 6. A department head has four tasks to be performed and three subordinates. The subordinates differ in efficiency. The estimates of the time, each subordinate would take to perform is given below in the matrix. How should be allocate the tasks, one to each man, so as to minimize the total man hour?

	Men		
	9	26	15
Task	13	27	6
	35	70	15
	18	30	20

Solution : Here the matrix is not square. Therefore, we add one fictitious man to make it square.

Now, we solve the problem with the matrix.

	Men →	1	2	3	4
Task	*I*	9	26	15	0
↓	*II*	13	27	6	0
	III	25	20	15	0
	IV	18	30	20	0

Step I :— Subtract the smallest element of each row from its row. In this case the matrix will not be changed as in each row the minimum element is zero.

Now, subtract the smallest element of each column from its column. The matrixis

	1	2	3	4
I	0	6	9	0
II	4	7	0	0
III	26	0	9	0
IV	9	10	14	0

First we search for rows containing exactly one zero. The 4th row contains zero at (4, 4). Make assignment ❑ there and draw a vertical through it. Now turning to uncrossed columns. we see that first column contains zero at (1, 1); 2nd column contains zero at (2, 3) and 3rd column at (2, 2). Make assignment ❑ there and draw horizontal lines through them (shown in the figure).

I	[0]	... 6 ...	9	0
II	... 4 ...	... 7 ...	[0] ...	0
III	... 26 ...	[0]	... 9 ...	0
IV	9	10	14	[0]

Thus drawing four lines (equal to the order of the matrix) all the zeros are covered. Also, every row and every column contains one assignment is as follows.

I Job → 1 man   *II* Job → 3 man   *III* Job → 2 man

The total time (man hours) is 9 + 6 + 20 = 35.

Example 7. Solve the assignment problem represented by the following matrix.

	I	*II*	*III*	*IV*	*V*	*VI*
A	9	22	58	11	19	27
B	43	78	72	50	63	48
C	41	28	91	37	43	33
D	74	42	27	49	39	32
E	36	11	57	22	25	18
F	3	56	53	31	17	28

[Meerut O. R. 69, 72, Kanpur 69]

Solution : *Step I.* Subtracting the minimum element of each from its row, the matrix reduces to.

	I	*II*	*III*	*IV*	*V*	*VI*
A	0	13	49	2	10	18
B	0	35	29	7	20	5
C	13	0	63	9	17	5
D	47	15	0	22	12	5
E	25	0	46	11	14	7
F	0	53	50	28	14	25

Subtracting the minimum element of each column from its column, the matrix reduces to.

	I	*II*	*III*	*IV*	*V*	*VI*
A	0	13	49	0	0	13
B	0	35	29	5	10	0
C	13	0	63	7	7	0
D	47	15	0	20	2	0
E	25	0	46	9	4	2
F	0	53	50	26	4	20

Now we give the zero assignment in our usual manner and get the following matrix

	I	*II*	*III*	*IV*	*V*	*VI*
A	... 0 ...	... 13 ...	... 49 ...	$\boxed{0}$ ...	... 0 ...	... 13 ...
B	0	35	29	5	10	$\boxed{0}$
C	13	0	63	7	7	0
D	... 47 ...	... 15 ...	$\boxed{0}$	20 ...	... 2 ...	... 0 ...
E	25	$\boxed{0}$	46	9	4	2
F	$\boxed{0}$	53	50	26	4	20

Since row 3 and column 5 have got to assignment. Therefore we proceed to the next step.
Now the smallest of the element that do not have a line through this is 4. Subtracting this element 4 from all the elements that do not have a line through them and adding to every element that lies at the intersection of two lines and leaving the remaining unchanged we get the following matrix

	I	II	III	IV	V	VI
A	4	17	49	[0]	0	17
B	0	35	25	1	6	[0]
C	13	[0]	59	3	3	0
D	51	19	[0]	20	2	4
E	25	0	42	5	[0]	2
F	[0]	53	46	22	0	20

Thus the optimal assignment is

$A \to IV$
$B \to VI$
$C \to II$
$D \to III$
$E \to V$
$F \to I$

Thus minimum cost $Z = 11 + 48 + 28 + 27 + 25 + 3$, by consulting the matrix = 142

Another optimal solution of this assignment problem is as follows

$A \to IV$
$B \to I$
$C \to VI$
$D \to III$
$E \to II$
$F \to V$

and minimum cost $Z = 11 + 48 + 33 + 27 + 11 + 17$

$= 142$ (from given cost matrix)

Example 8. A company has 5 jobs to be done. The following matrix shows the return in Rs. of assigning ith (i = 1, 2, 3, 4, 5) machine to the jth job (j = 1, 2, 3, 4, 5). Again the five jobs to the five machines so as to maximize the total return.

Machine / job	1	2	3	4	5
1	5	11	10	12	4
2	2	4	6	3	5
3	3	12	5	14	6
4	6	14	4	11	7
5	7	9	8	12	5

Solution : Here, instead of minimizing the cost, we are to maximize the profit, In such cases we from a new matrix say, $B = [c'_{ij}]$ from the given matrix and the apply the usual procedure to this matrix. Matrix B is obtained in the following manner :

$$c¢_{ij} = \text{Max}\ (c_{ij} — c_{ij}), \quad i = 1, 2, ¼¼\ m, \quad j = 1, 2, ¼¼\ m$$

Note that (Max. c_{ij}) is the greatest element of the given matrix. Matrix B will contain atleast one zero element)

In this example the number (Max, c_{ij}) is 14. Hence, the new matrix is obtained by subtracting each element from 14.

Job / Machine	1	2	3	4	5
1	9	3	4	2	10
2	12	10	8	11	9
3	11	2	9	0	8
4	8	0	10	3	7
5	7	5	6	2	9

Step I : Subtract the minimum element of each row from its row. Then matrix reduces to

	1	2	3	4	5
1	7	1	2	0	8
2	4	2	0	3	1
3	11	2	9	0	8
4	8	0	10	3	7
5	5	3	4	0	7

Step II : Subtract the minimum element of each column from its column. The matrix obtained is

Table *Cont.*

	1	2	3	4	5
1	3	1	2	0	7
2	0	2	0	3	0
3	7	2	9	0	7
4	4	0	10	3	6
5	1	3	4	0	6

Step III : Rows 1 and 4 contain exactly one zero at (1, 4) and (4, 2) respectively. Make assignment [0] there and cross over all the zero of the corresponding column (or draw vertical lines through them). Column 1 contains only one zero at (2, 1) Cross-out all the zeros of the row in which it occur (or draw a horizontal line through [0] at (2, 1).

Lines drawn are shown in the following matrix

	1	2	3	4	5
1	3	1	2	[0]	7
2	[0]	... 2 ...	0	... 3 ...	... 0 ...
3	7	2	9	0	7
4	4	[0]	10	3	6
5	1	3	4	0	6

Clearly, the three lines are passing through all the zeros. Since the number of lines is not equal to the order of the matrix, the optimal solution has not been reached so we proceed to next step.

Step IV : The smallest element out of those which do not lie on any line is 1. Subtract it from all these elements which do not lie on any line and add it to all the elements placed at the intersection of lines and leaving the element crossed by one line unaltered. The matrix is

	1	2	3	4	5
1	2	1	1	[0]	6
2	0	3	[0]	4	0
3	0	2	8	0	6
4	3	[0]	9	3	5
5	[0]	3	3	0	5

But still the optimal assignment has not been reached. the minimum of uncrossed elements is 1. Repeat the same procedure.

	1	2	3	4	5
1	2	1	[0]	0	5
2	1	4	0	5	[0]
3	6	2	7	[0]	5
4	3	[0]	8	3	4
5	[0]	3	2	0	4

Clearly in the above matrix, each row and each column contains the marked zero [0] . Hence the optimal assignment as been reached. The details of this programme are as follows

Machine 1 → Job 3 Profit 10

Machine 2 → Job 5 Profit 5

Machine 3 → Job 4 Profit 14

Machine 4 → Job 2 Profit 14

Machine 5 → Job 1 Profit 7

Total profit according to this plan is

$$10 + 5 + 14 + 14 + 7 = \text{Rs. } 50.00$$

Example 9. A company has four territories open and four salesman available for assignment. the territories are not equally rich in their sales potential; it is estimated that a typical salesman operating in each territory would bring in the following annual sales

Territory *I*	:	Rs. 60,000
" *II*	:	Rs. 50,000
" *III*	:	Rs. 40, 000
" *IV*	:	Rs. 30,000

The four salesmen are considered to differ in ability, it is estimated that working under the same condition, the yearly sales would be proportionately as follows—

Salesman	A	B	C	D
Proportion	7	5	5	4

If criterion is maximum expected total sales the intuitive answer is to assign the best salesman to the richest territory and so on. Verify this answer by the assignment method.

[Gauhaty M. Sc. Stat. 71, Kanpur 73, Meerut Stat 71]

Solution : The sum of proportions of four salesman = 7 + 5 + 5 + 4 = 21. If we consider the sale of Rs. 10,000 as one unit of sale then annual sale is the four territories are follows

$$\text{For } A, \frac{7}{21} \times 6, \frac{7}{22} \times 5, \frac{7}{21} \times 4, \frac{7}{21} \times 3$$

$$\text{For } B, \frac{5}{21} \times 6, \frac{5}{21} \times 5, \frac{5}{21} \times 4, \frac{5}{21} \times 3$$

$$\text{For } C, \frac{5}{21} \times 6, \frac{5}{21} \times 5, \frac{5}{21} \times 4, \frac{5}{21} \times 3$$

$$\text{For } D, \frac{4}{21} \times 6, \frac{4}{21} \times 5, \frac{4}{21} \times 4, \frac{4}{21} \times 3$$

If we determine the assignments which make the total annual sales maximum, then the effectiveness matrix will be as follows

		6	5	4	3 ← Sales in Rs.10.000
Sales proportion		*I*	*II*	*III*	*IV*
7	*A*	$\frac{42}{21}$	$\frac{35}{21}$	$\frac{28}{21}$	$\frac{21}{21}$
5	*B*	$\frac{30}{21}$	$\frac{25}{21}$	$\frac{20}{21}$	$\frac{15}{21}$
5	*C*	$\frac{30}{21}$	$\frac{25}{21}$	$\frac{20}{21}$	$\frac{15}{21}$
4	*D*	$\frac{24}{21}$	$\frac{20}{21}$	$\frac{16}{21}$	$\frac{12}{21}$

Consider the sale of 21 years in place of annual sales in order to avoid the fraction. The sales in 21 years are as follows

For *A* : 42, 35, 28, 21

For *B* : 30, 25, 20, 15

For *C* : 30, 25, 20, 15

For *D* : 24, 20, 16, 12

Thus the problem is to determine the assignment of (salesman to territories) which make the total sales maximum and the effective matrix is

	I	*II*	*III*	*IV*
A	42	35	28	21
B	30	25	20	15
C	30	25	20	15
D	24	20	16	12

This is a maximization problem. To convert into a minimization problem, we form a new matrix by multiplying each element of this matrix by (–1), The resulting matrix for the minimization problem is

	I	*II*	*III*	*IV*
A	– 42	– 35	– 28	– 21
B	– 30	– 25	– 20	– 15
C	– 30	– 25	– 20	– 15
D	– 24	– 20	– 16	– 12

Step I : Subtracting the minimum element of each row from that row, we get

	I	II	III	IV
A	0	7	14	21
B	0	5	10	15
C	0	5	10	15
D	0	4	8	12

Subtracting the smallest element of each column from that column, we have

	I	II	III	IV
A	0	3	6	9
B	0	1	2	3
C	0	1	2	3
D	0	0	0	0

Step II : We note that the minimum number of lines which cross all the zeros is 2. Thus a complete assignment is not possible at this iterations.

Step III : Smallest of all uncrossed elements is 1. Subtract this elements from all unmarked elements and add it to the elements placed at the intersection. The new matrix is

	I	II	III	IV
A	0	2	5	8
B	0	0	1	2
C	0	0	1	2
D	1	0	0	0

Still we do not get a complete assignment. Again, apply the step III.

We get the matrix as shown below

	I	*II*	*III*	*IV*
A	$\boxed{0}$	2	4	7
B	0	$\boxed{0}$	0	1
C	0	0	$\boxed{0}$	1
D	2	1	0	$\boxed{0}$

The optimal solutions are

$$\text{(i)} \quad A \to I, B \to III, \quad C \to II, \quad D \to IV$$

The other solution

$$\text{(ii)} \quad A \to I, B \to II, \; C \to III, \quad D \to IV$$

From both the solutions it is obvious that the best salesman A is assignment to the richest territory I and the worst salesman D to the poorest territory D. The salesman B and C are equal in efficiency, so either of them may be assigned to territory II and III. This assignment verify the given problem

10.Solve the following assignment problem.

	I	*II*	*III*	*IV*	*V*
A_1	∞	4	7	3	4
A_2	4	∞	6	3	4
A_3	7	6	∞	7	5
A_4	3	3	7	∞	7
A_5	4	4	5	7	∞

Solution : Subtracting the smallest element of each row from that row, we get

	I	*II*	*II*	*IV*	*V*
A_1	∞	1	4	0	1
A_2	1	∞	3	0	1
A_3	2	1	∞	2	0
A_4	0	0	4	∞	4
A_5	0	0	1	3	∞

Now, subtract the minimum element of each column from its column.

	I	II	III	IV	V
A_1	∞	1	3	0	1
A_2	1	∞	2	0	1
A_3	2	1	∞	2	0
A_4	0	0	3	∞	4
A_5	0	0	0	3	∞

Give the zero assignments in usual manner and get the following matrix

	A_1	A_2	A_3	A_4	A_5
A_1	∞	1	3	[0]	1
A_2	1	∞	2	0	1
A_3	2	1	∞	2	[0]
A_4	[0]	0	3	∞	4
A_5	0	[0]	0	3	∞

Obviously, this does not give optimal assignment. So we proceed to the next step.

Employing the procedure of assignment algorithm, we

	I	II	III	IV	V
A_1	∞	[0]	2	0	1
A_2	0	∞	1	[0]	1
A_3	1	0	∞	2	[0]
A_4	[0]	0	3	∞	5
A_5	0	0	[0]	4	∞

The optimal assignment is

$$A_1 \to II, A_2 \to IV, \quad A_3 \to V, \quad A_4 \to I, A_5 \to III$$

Example 11. Average time taken by an operator on a specific machine is tabulated below. The management is considering to replace one of the old machines by a new one and the estimated time for operation by each operator on the new machine is also indicated.

Machine

Operators	1	2	3	4	5	6		New
	A	10	12	8	10	8	12	11
	B	9	10	8	7	8	9	10
	C	8	7	8	8	8	6	8
	D	12	13	14	14	15	14	11
	E	9	9	9	8	8	10	9
	F	7	8	9	9	9	8	8

(a) Find out an allocation of operators to the old machines to achieve a minimum operator time.

(b) Reset the problem with new machine and find out the allocation of the operation to each machine and comment whether it is advantageous to replace an old machine to achieve a reduction in operating time only.

(c) How all the operators be reallocated to the machines after replacement.

[I. C. W. A; Final (N. S. Jue 78]

Solution : Firstly, we solve the problem of assigning 6 operators *A*, *B*, *C*, *D*, *E*, *F* to six machines 1, 2, 3, 4, 5, 6.

This assignment problem can be solved by usual Assignment Algorithm. The optimal solutions are

$$A \to 3$$
$$B \to 4$$
$$C \to 6$$
$$D \to 1$$
$$E \to 5$$
$$F \to 2$$

or

$$A \to 3$$
$$B \to 4$$
$$C \to 6$$
$$D \to 2$$
$$E \to 5$$
$$F \to 1$$

Minimum time

$$= 8 + 7 + 6 + 12 + 8 + 8$$
$$= 49 \text{ hours}$$

(b) Now, reset the problem in the following way by introducing the new machine also.

Operators	Machine 1	2	3	4	5	6	New
A	10	12	8	10	8	12	11
B	9	10	8	7	8	9	10
C	8	7	8	8	8	6	8
D	12	13	14	14	15	14	11
E	9	9	9	8	8	0	9
F	7	8	9	9	9	8	8

It is an unbalanced assignment problem. Solve this by introducing a dummy operator. The new matrix is now.

	1	2	3	4	5	6	New
A	10	12	8	10	8	12	11
B	9	10	8	7	8	9	10
C	8	7	8	8	8	6	8
D	12	13	14	14	15	14	11
E	9	9	9	8	8	10	9
F	7	8	9	9	9	8	8
G	0	0	0	0	0	0	0

Adopting the usual procedure we get

$$A \to 3, B \to 4, C \to 6, D \to \text{New}, \quad E \to 5, \quad F \to 1$$

Minimum time = 8 + 7 + 6 + 11 + 8 + 7 = 47 hours

This shows that if new machine is introduced in place of 2nd old machine, the operator time can be reduced by 2 hours. But the operation costs of these machines are not given. so we cannot say whether the replacement of 2nd old machine by a new machine will be advantageous or will merely decrease the operative time by 2 hours.

However from the given data it is clear that operator *D* takes maximum time to operate any of the machines. It means that *D* is not appropriately trained hence he should be given appropriate training to operate the machine so that he may bring his operative times to the level of atleast *A*.

(c) After replacing machine 2 by a new one, the optimal assignments operators are as follows—

$A \to 3$
$B \to 4$
$C \to 6$
$D \to \text{New}$
$E \to 5$
$F \to 1$

Total minimum operator cost = 47 hours.

Example 12. Solve the minimal assignment problem

Man →		A	B	C	D	E
Job ↓	*1*	2	3	7	15	14
	II	7	9	17	14	10
	III	9	6	12	6	7
	IV	7	6	14	6	10
	V	9	6	12	10	6

[Delhi 65, Indian Statistical Institute May 74]

Solution : Subtract the smallest element of each row from every element of the corresponding row.0

Man →		A	B	C	D	E
Job ↓	*I*	0	6	5	13	12
	II	0	2	10	7	3
	III	3	0	6	0	1
	IV	1	0	8	0	10
	V	3	0	6	4	0

Now, Subtract the smallest element of each column from every element of the corresponding column.

	A	B	C	D	E
I	0	2	$\boxed{0}$	13	12
II	$\boxed{0}$	3	5	7	3
III	3	0	1	$\boxed{0}$	1
IV	1	$\boxed{0}$	3	0	10
V	3	0	1	4	$\boxed{0}$

Example 13. Solve the following minimal assignment problem

Machine ↓ \ Job →	A_1	B_1	C_1	D_1	E_1
A	5	11	10	12	4
B	2	4	6	3	5
C	3	12	5	14	6
D	6	14	4	11	7
E	7	9	8	12	5

[R. U. Condensed Course Hons. 88 H, 90 H, Bombay B. Sc. Stat. 74]

Solution : Subtract the smallest element of each row from corresponding row

	A_1	B_1	C_1	D_1	E_1
A	1	7	6	8	0
B	0	2	4	1	5
C	0	9	2	11	4
D	2	10	0	7	3
E	2	4	3	7	0

Now, subtract the smallest element of each column from every element of the corresponding column.

	A_1	B_1	C_1	D_1	E_1
A	1	5	6	7	$\boxed{0}$
B	... 0 ...	$\boxed{0}$	... 4 ...	... 0 ...	... 5
C	$\boxed{0}$	7	2	10	4
D	2	8	$\boxed{0}$	6	3
E	2	2	3	6	0

The smallest element which does not lie on ay line is 2. Subtract this from all the elements which does not lie on any line and add to those elements which lie on the intersection.

	A_1	B_1	C_1	D_1	E_1
A	1	3	6	5	[0]
B	2	0	6	[0]	7
C	[0]	5	2	8	4
D	2	8	[0]	4	3
E	2	[0]	3	4	0

$$A \to E_1$$
$$B \to D_1$$
$$C \to A_1$$
$$D \to C_1$$
$$E \to B_4$$

Minimum cost = 4 + 3 + 3 + 4 + 9 = 23

Example 14. Find the minimum cost solution for 5 × 5 assignment problem where costs are as given below—

	I	*II*	*III*	*IV*	*V*
1	—2	—4	—8	—6	—1
2	0	—9	—5	—5	—4
3	—3	—8	—9	—2	—6
4	—4	—3	—1	—0	—3
5	—9	—5	—8	—9	—3

[Roorkee M. E. (Meerut 77 Kharagpur 78]

Solution : Subtract the smallest element of each row from the corresponding row—

	I	*II*	*III*	*IV*	*V*
1	6	4	0	2	7
2	9	0	4	4	5
3	6	1	0	7	3
4	0	1	3	4	1
5	0	4	1	0	6

Subtract the smallest element of each column from the corresponding column.

	I	II	III	IV	V
1	6	4	[0]	2	6
2	9	[0]	4	4	4
3	6	1	0	7	2
4	[0]	… 1 …	… 3 …	… 4 …	[0]
5	0 …	… 4 …	… 1 …	[0]	5 …

	I	II	III	IV	V
1	6	4	[0]	2	6
2	9	[0]	4	4	4
3	6	1	0		2
4	[0]	… 1 …	… 3 …	… 4 …	… 0 …
5	0 …	…4 …	… 1 …	[0]	… 5 …

The third row (or Vth column) has got no assignment so we adopt the following method.

The smallest element which does not lie on any lie is 2. Subtract 2 from each element which does not lie on any line and add it to those which lie on the intersection..

	I	II	III	IV	V
1	6	4	[0]	0	4
2	9	[0]	4	2	2
3	6	1	0	5	[0]
4	[0]	3	5	4	0
5	0	6	3	[0]	5

$1 \to III$, $2 \to II$, $3 \to V$

$4 \to I$, $5 \to IV$

Minimum cost = 36

EXERCISE 5

1. Given an algorithm to solve an assignment problem.
[Kanpur 69, 73]

2. Write a short note on :Assignment Problem".
[V. B. U. Part III Hons. 99, Meerut O. r. 74 S, 78; Kanpur 69]

3. Describe the Hungarian method of solving the assignment problem.
[Sardar Patel Univ. Stat. 75]

4. Solve the following minimal assignment problems.

(i)

Man → / Job ↓	I	II	III	IV	V
A	1	3	2	3	6
B	2	4	3	1	5
C	5	6	3	4	6
D	3	1	4	2	2
E	1	5	6	5	4

[Meerut Stat. 74]

Ans. $A \to I$, $B \to IV$, $C \to III$, $D \to II$, $E \to V$

(ii)

Man → / Job ↓	1	2	3	4
I	12	30	21	15
II	18	33	9	31
III	44	25	24	21
IV	23	30	28	14

[Delhi 87]

Ans. $I \to I$, $II \to 3$, $III \to 2$, $IV \to 4$.

(iii)

Man → / Job ↓	1	2	3	4	5
I	2	8	7	15	14
II	7	9	17	14	10
III	9	6	12	6	7
IV	7	6	14	6	10
V	9	6	12	10	6

[Delhi 65, Indian Statistical Institute May 74]

Ans. $I \to 3$, $II \to 1$, $III \to 2$, $IV \to 4$, $V \to 5$.

(iv)

Man → / Job ↓	*I*	*II*	*III*	*IV*
1	2	3	4	5
2	4	5	6	7
3	7	8	9	8
4	3	5	8	4

[Delhi 69]

Ans. $1 \to I$, $2 \to II$, $3 \to III$, $4 \to IV$.
Or, $1 \to II$, $2 \to I$, $3 \to IV$, $4 \to III$.

5. Solve the following Assignment problem.

Project ↓ / Location	*I*	*II*	*III*	*IV*	*V*
A	15	21	6	4	9
B	3	40	21	0	7
C	9	6	5	8	10
D	14	8	6	9	3
E	21	1	18	7	4

[Gauhati Statistics 69]

Ans. $A \to IV$, $B \to I$, $C \to II$, $D \to III$, $E \to V$.

6. Solve the following assignment problem.

	1	2	3	4	5	6
I	9	22	58	11	9	27
II	43	78	72	50	63	48
III	41	28	91	37	45	33
IV	74	42	27	49	39	32
V	36	11	57	22	25	18
VI	3	56	53	31	17	28

[Meerut Math 69, 72, Kanpur 69]

$I \to 4$, $II \to 1, III \to 6$, $IV \to 3$, $V \to 2$, $VI \to 5$

7. Solve the following assignment problem.

	1	2	3	4
A	10	12	19	11
B	5	10	7	8
C	12	14	13	11
D	8	15	11	9

Ans. $A \to 2$, $B \to 3$, $C \to 4$, $D \to 1$
Minimum cost Z = 38

[Shivaji M. Sc. Math 76]

8. Solve the following minimal assignment problem.

Job → Machine	*I*	*II*	*III*	*IV*	*V*
A	5	11	10	12	4
B	2	4	6	3	5
C	3	12	5	14	6
D	6	14	4	11	7
E	7	9	8	12	5

[Bombay B. Sc. Stat. 74, B. Sc. Spl. Hons. 88 H, R. U. 90 H]

Ans. $A \to V$, $B \to IV$, $C \to$ I, $D \to III$, $E \to II$

Minimum cost Z = 23.

9. Find the minimum cost solution for the 5 x 5 assignment problem where coefficient are as given below.

	I	*II*	*III*	*IV*	*V*
1	—2	4	—8	—6	—1
2	0	—9	—5	—5	—4
3	—3	—8	—9	—2	—6
4	—4	—3	—1	—0	—3
5	—9	—5	—8	—9	—3

[Roorkee M. E. (Mech.) 77, Kharagpur 78]

Ans. $1 \to III$, $2 \to II$, $3 \to V$, $4 \to I$, $5 \to IV$.
or I $\to IV$, $2 \to III$, $3 \to II$, $4 \to V$, $5 \to I$.
Cost = 36

10. A marketing manager has 5 salesman and 5 sales districts. Considering the capabilities of the salesman and the nature of districts, the marketing manager estimates that sales per month (in hundred rupees) for each salesman in each districts would be as follows—

Job	*A*	*B*	*C*	*D*	*E*
1	32	38	40	28	40
2	40	24	28	21	36
3	41	27	33	30	37
4	22	38	41	36	36
5	29	33	40	35	39

Find the assignment of salesman to districts that will result in maximum sales.

[Delhi M. B. A. 75. 77]

Ans. $1 \to B$, $2 \to A$, $3 \to E$, $4 \to C$, $5 \to D$
Maximum $Z = 191$

11. Five men are available to do five different jobs from past records, the time (in hours) that each man takes to do each hob is known and is given in the following table.

Man	*I*	*II*	*III*	*IV*	*V*
1	2	9	2	7	1
2	6	8	7	6	1
3	4	6	5	3	1
4	4	2	7	3	1
5	5	3	9	5	1

Find the assignment of men to jobs that will minimize the time taken.

Ans. $1 \to III$, $2 \to V$. $3 \to I$, $4 \to IV$, $5 \to II$
or $1 \to III$, $2 \to V$, $3 \to IV$, $4 \to II$, $5 \to I$
or $1 \to III$, $2 \to V$, $3 \to IV$, $4 \to I$, $5 \to II$

12. One car is available at each stations 1, 2, 3, 4, 5, 6 : and one car is required at each of the stations 7, 8, 9, 10, 11, 12. The distances between the various stations are given in the matrix below. How should the cars be dispatched so as to minimize the total mileage traveled.

	7	8	9	10	11	12
1	42	72	39	52	25	51
2	22	29	49	65	81	50
3	27	39	60	51	32	32
4	45	50	48	52	37	43
5	29	40	39	26	30	33
6	82	40	40	60	51	30

[Delhi 66]

Ans. $1 \to 11$, $2 \to 8$, $3 \to 12$, $4 \to 9$, $5 \to 10$, $6 \to 7$
Minimum mileage $Z = 185$.

6

Theory of Games

6.1. SOME IMPORTANT DEFINITIONS

1. Games—Game is defined as activity between two or more persons involving activities by each person according to a set of rules, at the end of which each person gets some benefit or suffers loss.
2. Play—The set of rules defines the game. Going through the set of rules once by the participants defines a play.
3. Chance of strategy—If in a game, activities are determined by skill, it is said to be game of strategy; if they are determined by chance, it is a **Game of Chance.**

   [R. U. Part II Hons. 91]
4. Two-person game—Suppose there are finite number (n) of players such that $n \geq 2$. In case $n = 2$, it is called a **Two-person game.**
5. Strategy—(Meerut 73 (S), 83, V. B. U. Part III Hons. 97)

   A strategy for a given player is a set of rules that specifies which of the available courses of action he should make at each play.

There are two types of strategy

(i) Pure strategy (Kuruk 79, keral (Stat) 83, Agara 86, Bhopal (Stat) 86, Meerut 74 (S), 75, 83, Rohilkhand 85, V. B. U. Part III Hons. 99 (Re-exam.)

If a player knows exactly what the other player is going to do, a deterministic situation is obtained and the objective function is to maximize the gain.

Thus, the pure strategy is a decision rule always to select a particular course of action.

(ii) Mixed strategy → (Kuruk 75, Keral (Stat) 83, Meerut 74 (S), 75, 83, Bhopal (Stat) 86, Agara 86, Rohilkhand 85, V. B. U. Part Hons. III 99 (Re-exam.)

If a player is guessing as to which activity is to be selected by the other on any particular occasion, a Probabilistic Situation is obtained and objective function is to maximize the expected gain.

The mixed strategy is a selection among pure strategies with fixed probability.

6. Two person Zero-Sun Games (Rectangular Games)

[Calcuta 76, Kuruk 79, Meerut 70, 71, 72, 77, 83, R. U. Part III Hons. 91, V. B. U. Part III Hons. 96, 97, 98]

A game with only two players (say, A and B) is called a "Two-person, zero-sum game" if the losses of one player are equivalent to the gains of other, so that the sum of their net gains is zero. Two-person, zero-sum-games are called Rectangular games as these are generally represented by a Pay-off-matrix in rectangular form.

7. Pay-off matrix (or gain matrix)

[V. B. U. Part III Hons, 96, 98]

In a two-person zero-sum game, the resulting gain is represented in the form of matrix, called the Pay-off-matrix.

Thus, a pay-off-matrix is a table which shows how payment should be made at the end of a play.

If the player A has m activities and B has n activities. Then A's pay-off-matrix is the cell entry is

				B			
		1	2	3	j	…	n
	1	a_{11}	a_{12}	a_{13}	a_{ij}	…	a_{1n}
	2	a_{21}	a_{22}	a_{23}	a_{2j}	…	a_{2n}
A	3	a_{31}	a_{32}	a_{33}	a_{3j}	…	a_{3n}
	⋮	⋮	⋮	⋮	⋮		⋮
	i	a_{i1}	a_{i2}	a_{i3}	a_{ij}	…	a_{in}
	m	a_{m1}	a_{m2}	a_{m3}	a_{mj}	…	a_{mn}

The cell entry a_{ij} is the payment to A when A. Chooses the course of action i and B chooses the course of action j.

B' pay-off matrix

		B				
		1	2	3	....	n
	1	$-a_{11}$	$-a_{12}$	$-a_{13}$	....	a_{1n}
	2	$-a_{21}$	$-a_{22}$	$-a_{23}$	....	a_{2n}
A	3	$-a_{31}$	$-a_{32}$	$-a_{33}$	....	a_{3n}
	:	:	:	:		:
	m	$-a_{m1}$	$-a_{m2}$	$-a_{m3}$	....	a_{mn}

8. Saddle point—A saddle point is a position of such an element in the pay-off matrix, which is the minimum in its row and maximum in its column. The pay off at the saddle point is called the value of the game.
   A game may have more than one saddle point.

[R. U. Part III Hons. 91]

Theorem—For any 2×2 two person zero-game without any saddle point having pay-off matrix for player A is

		B	
		$B_1\ (q_1)$	$B_2\ (q_2)$
A	$A_1(p_1)$	a_{11}	a_{12}
	$A_2\ (p_2)$	a_{21}	a_{22}

the optimal strategies (p_1, p_2) and (q_1, q_2) are given by

$$\frac{p_1}{p_2} = \frac{a_{22} - a_{21}}{a_{11} - a_{12}} \text{ and } \frac{q_1}{q_2} = \frac{a_{22} - a_{21}}{a_{11} - a_{21}}$$

and the value of the game to A is given by

$$V = \frac{a_{11}\ a_{22} - a_{12}\ a_{21}}{(a_{11} + a_{22}) - (a_{11} + a_{21})}$$

[Meerut 77, 81, 82, 82 (P), 86, Agara 82, 87, Rohilkhand 88,Bhopal (Stat) 85, V. B. U. Part III Hons. 96, 98]

Proof—Let a mixed strategy for A be given by

$$S_A = \begin{bmatrix} A_1 & A_2 \\ P_1 & P_2 \end{bmatrix}$$

where $\quad P_1 + P_2 = 1.$

Thus if B moves B_1, the net expected gain of

A will be $\quad E_1\ (p) = a_{11}\ p_1 + a_{21}\ p_2.$

Again, if B moves B_2 , the net expected gain of A will be

$$E_2\ (p) = a_{12} p_1 + a_{22} p_2$$

Similarly, if B plays his mixed strategy

$$S_B = \begin{bmatrix} B_1 & B_2 \\ q_1 & q_2 \end{bmatrix}$$

where $q_1 + q_2 = 1$, then B's expected loss will be

$$E_1(q) = a_{11} q_1 + a_{12} q_2$$

If $\quad$ A plays A_1.
If $\quad$ A plays A_2, then

$$E_2(q) = a_{21} q_1 + a_{22} q_2$$

If V be the value of game, then we must have

$$E_1(p) \geq V, E_2(p) \geq V;$$

$$E_1(q) \leq V, E_2(q) \leq V.$$

When S_A and S_B happen to be optimum strategies, these inequalities become strict equations together with $p_1 + p_2 = 1$ and $q_1 + q_2 = 1$ can be solved for p_1 , p_2 and q_1 , q_2.

Solving $\quad$ $E_1(p) = V = E_2(p)$, we get

$$\frac{p_1}{p_2} = \frac{a_{22} - a_{21}}{a_{11} - a_{12}} \quad ...(i)$$

and solving $E_1(q) = E_2(q) = V$, we have

$$\frac{q_1}{q_2} = \frac{a_{22} - a_{21}}{a_{11} - a_{21}} \quad ...(ii)$$

Using $p_1 + p_2 = 1$, we get from (i),

$$p_1 = \frac{a_{22} - a_{21}}{(a_{11} + a_{22}) - (a_{12} + a_{21})}$$

$$p_2 = \frac{a_{22} - a_{12}}{(a_{11} + a_{22}) - (a_{12} + a_{21})}$$

Again, with the aid of $q_1 + q_2 = 1$, we get from (ii).

$$q_1 = \frac{a_{22} - a_{12}}{(a_{11} + a_{22}) - (a_{12} + a_{21})}$$

$$q_2 = \frac{a_{22} - a_{21}}{(a_{11} + a_{22}) - (a_{12} + a_{21})}$$

Putting $\quad$ p_1, p_2 in $E_1(p) = v$

i. e. in $\quad$ $a_{11} p_1 + a_{21} p_2 = v$, we get

$$v = \frac{a_{11}\, a_{22} - a_{21}\, a_{12}}{(a_{11} + a_{22}) - (a_{12} + a_{21})}$$

10. Rule of Dominance

[V. B. U. Part III Hons. 98, 99 (Re-exam.)]

(i) If all elements of a row, say i th, are less than or equal to the corresponding elements of any other row, say, j th, then i th row is dominated by j th row.

(ii) If all the elements of column say, ith, are greater than or equal to the corresponding elements of any other column, say, jth, then ith column is dominated by jth column.

(iii) A pure strategy may be dominated, if it is inferior to an average of two or more other pure stategies.

IIlustrative Examples

1. Solve the game whose pay off matrix is given by

		B		
		I	*II*	*III*
A	*I*	—2	15	—2
	II	—5	—6	—4
	III	—5	20	—8

[Kuruk 79, Rohilkhand '81]

Solution :

(i) Minimum elements of first row are –2, at (1, 1) and at (1, 3). Denote if by (– 2)

Minimum element of 2nd row is –6 which is shown as (– 6)

Minimum element of 3rd row is –8 shown as (– 8)

(ii) Maximum element of first column is –2, Denoted by [– 2]

Maximum element of 2nd column is 20 shown as [20]

Maximum element of 3rd column is –2 denoted by [– 2]

The saddle point is shown enclosed by a circle and a square both [O]

See the following table

		B I	B II	B III
A	I	−2	15	−2
A	II	−5	−6	−4
A	III	−5	20	−8

This game has two saddle points (1, 1) and (1,3).
Thus the solution to this game is given by

(i) the best strategy for A is I

(ii) the best strategy for B is either I or III

(iii) the value of the game is –2 for A and + 2 for B.

2. Solve the following game

		B I	B II	B III
A	I	6	8	6
A	II	4	12	2

[Meerut 77, 81, 82 (p) Jamia B. Sc. Math 77]

Solution : Circling O the row minima and putting squares □ around the column maxima, we get the following table.

		B I	B II	B III
A	I	6	8	6
A	II	4	12	2

The saddle points are shown by [O]
This game has two saddle points at (1, 1) and (1, 3)

(i) The best strategy for A is I
(ii) The best strategy for B is either I or III
(iii) The value of the game is 6 for A and -6 for B.

3. Solve the game whose any pay off matrix to player A

		B		
		I	*II*	*III*
	I	1	7	2
A	*II*	6	2	7
	III	0	1	6

[Meerut 72, 75, Bombay (M. Com.) 75, B. Sc. (Stat) 74, Shivajee 77, Delhi (M. Com.) 76, B. Sc. (Math) 77, M. Sc. OR, Delhi) 88, 89,V. B. U. Part III Hons. 99 (Re-exam.)]

Solution : Since the third row is inferior to the 2nd row, therefore 3rd row can be left out. Thus the reduced matrix is

		B		
		I	*II*	*III*
	I	1	7	2
A	*II*	6	2	7

Again, column III is dominated by column I, therefore column III can be deleted.

The reduced matrix is

		B	
		I	*II*
	I	1	7
A	*II*	6	2

Compare it with

		B	
		$I(q_1)$	$II\ (q_2)$
	I	1	7
A	*II*	6	2

Then $a_{11} = 1, a_{12} = 7, a_{21} = 6, a_{22} = 2$

$$p_1 = \frac{a_{22} - a_{21}}{(a_{11} + a_{22}) - (a_{12} + a_{21})} = \frac{2-6}{(1+2)-(7+6)} = \frac{2}{5}$$

$$p_2 = \frac{a_{11} - a_{12}}{(a_{11} + a_{22}) - (a_{12} + a_{21})} = \frac{1-7}{(1+2)-(7+6)} = \frac{3}{5}$$

$$q_1 = \frac{a_{22} - a_{12}}{(a_{11} + a_{22}) - (a_{12} + a_{21})} = \frac{1}{2}$$

$$q_2 = \frac{a_{11} - a_{21}}{(a_{11} + a_{22}) - (a_{12} + a_{21})} = \frac{1}{2}$$

$$v = \frac{a_{11} \times a_{22} - a_{12} \times a_{21}}{(a_{11} + a_{22}) - (a_{12} + a_{21})} = \frac{1 \times 2 - 7 \times 6}{(1+2) - (7+6)} = 4$$

Thus, the following solution is obtained

(i) A chooses optimal strategies $(p_1, p_2, p_3) = \left(\frac{2}{5}, \frac{3}{5}, 0\right)$

(ii) B chooses optimal strategies $(q_1, q_2, q_3) = \left(\frac{1}{2}, \frac{1}{2}, 0\right)$

The value of the game for A is $v = 4$.

5. Solve the game whose pay off matrix is given by

B

		I	*II*	*III*	*IV*
	I	3	2	4	0
	II	2	4	2	4
A	*III*	4	2	4	0
	IV	0	4	0	8

[Calcutta 77, Kuruk 75, Meerut 70, 71, 75, 79, 84, 85, 88,Delhi (B. Sc. Math) 75, 77, 78, Rohilkhand 83, 88, Agara 89, M. Com. 76, V. B. U. Part III Hons. 98]

Solution :

This game has no saddle point.

First row is dominated by 3rd row from A's point of view. So delete First row.

The reduced matrix is

B

		I	*II*	*III*	*IV*
	II	2	4	2	4
A	*III*	4	2	4	0
	IV	0	4	0	8

From B's Point of view, first column is dominated by 3rd column so the reduced matrix is

		B		
		II	*III*	*IV*
A	*II*	4	2	4
	III	2	4	0
	IV	4	0	8

The average of the *B*'s III and IV pure strategies are

$$\left\{\frac{2+4}{2}, \frac{4+0}{2}, \frac{0+8}{2}\right\} = \{3, 2, 4\}$$

which is clearly superior to B's II pure strategies, under the condition B will not choose II strategies.

Hence column II may be deleted.

Thus the new matrix is

		B	
		III	*IV*
A	*II*	2	4
	III	4	0
	IV	0	8

In the new matrix, the average of A's III and IV pure strategies given $\left\{\frac{4+0}{2}, \frac{0+8}{2}\right\} = \{2, 4\}$

Which is obviously the same as A's II startegy.

Hence, delete A's II startegy.

The reduced (2×2) matrix is

		B	
		III (q_1)	*IV* (q_2)
A	*III* (p_1)	4	0
	IV (p_2)	0	8

Here $a_{11} = 4, a_{12} = 0$

$a_{21} = 0, a_{22} = 8$

$$p_1 = \frac{a_{22} - a_{21}}{(a_{11} + a_{22}) - (a_{12} + a_{21})} = \frac{8 - 0}{(4+8) - (0+0)} = \frac{8}{12} = \frac{2}{3}$$

$$p_2 = 1 - \frac{2}{3} = \frac{1}{3} \quad \text{as } p_1 + p_2 = 1$$

$$q_1 = \frac{a_{22} - a_{12}}{(a_{11} + a_{22}) - (a_{12} + a_{21})} = \frac{2}{3}$$

$$q_2 = 1 - q_1 = 1 - \frac{2}{3} = \frac{1}{3} \text{ as } q_1 + q_2 = 1$$

v = value of the game

$$= \frac{a_{11} \times a_{22} - a_{11} \times a_{21}}{(a_{11} + a_{22}) - (a_{12} + a_{21})} = \frac{8}{3}$$

Hence the solution is

(i) Optimal strategy for A is $\left(0, 0, \frac{2}{3}, \frac{1}{3}\right)$

(ii) Optimal strategy for B is $\left(0, 0, \frac{2}{3}, \frac{1}{3}\right)$

(iii) Value of the game to A is $\frac{8}{3}$

6. Solve the game whose pay-off-matrix is

		B: *I*	*II*	*III*
A	*I*	–1	–2	8
	II	7	5	–1
	III	6	0	12

[Meerut 74 (S), 75 (S), R. U. 88 H, R. U. Part III Hons. 91, V. B. U. Part III Hons. 97 H]

Solution :

The game has no saddle point. Every element of 3rd row is greater than the corresponding element of first row. So from A's point of view startegy I is dominated by III. So delete Ist row. The reduced matrix is.

		B: *I*	*II*	*III*
A	*II*	7	5	–1
	III	6	0	12

In this new matrix, every element of first column is greater than the corresponding element of 2nd column.

So from B's point of view strategy I is dominated by strategy I, So delete first column.

The reduced (2 × 2) matrix is

		B	
		$II(q_1)$	$III\ (q_2)$
	$II\ (p_1)$	5	−1
A	$III\ (p_2)$	0	12

Let A choose his II and III strategies with probabilities p_1 , p_2 and let B choose his II and III with probabilitis q_1, q_2 respectively.

Here $a_{11} = 5,\quad a_{12} = -1$

$a_{21} = 0,\quad a_{22} = 12$

$$p_1 = \frac{a_{22} - a_{21}}{(a_{11} + a_{22}) - (a_{12} + a_{21})} = \frac{2}{3}$$

$$p_2 = 1 - p_1 = 1 - \frac{2}{3} = \frac{1}{3} \text{ as } p_1 + p_2 = 1$$

$$q_1 = \frac{a_{22} - a_{12}}{(a_{11} + a_{22}) - (a_{12} + a_{21})} = \frac{13}{18}$$

$$q_2 = 1 - \frac{13}{18} = \frac{5}{18} \text{ as } q_1 + q_2 = 1$$

v = value of the game

$$= \frac{a_{11}\, a_{22} - a_{12}\, a_{21}}{(a_{11} + a_{22}) - (a_{12} + a_{21})} = \frac{10}{3}$$

So the solution of game is

(i) optimal strategy for A is $\left(0, \frac{2}{3}, \frac{1}{3}\right)$

(ii) optimal strategy for B is $\left(0, \frac{13}{18}, \frac{5}{18}\right)$

(iii) value of game for a is $\frac{10}{3}$.

7. Solve the game

		B		
		I	II	III
	I	−3	−2	6
A	II	2	0	2
	III	5	−2	−4

[R. U. 90 H Condensed course)

Solution :

(i) Encircle the minima of row by O.

(ii) Put maxima of column in □ .

(iii) The saddle point is [O]

The above table can be shown as

		B I	B II	B III
	I	(−3)	−2	[6]
A	II	2	[(0)]	2
	III	[5]	−2	(−4)

Solution :

The solution of game is

(i) The best strategy for A is II

(ii) The best strategy for B is II

(iii) The value of same is $v = 0$

8. The pay off matrix of a game is below. Find the best strategy for each player, and the value of a play of the game to A and B.

		B I	B II	B III	B IV	B V
	I	9	3	1	8	0
	II	6	5	4	6	7
A	III	2	4	3	3	8
	IV	5	6	2	2	1

[R.U. Part III Hons. 91]

Solution :

(i) Encircle the minima of row by O

(ii) Put the maxima of the columns in □

(iii) Show the saddle point by [O]

Then above table can be shown as

	I	9 (boxed)	3	1	8 (boxed)	0 (circled)
	II	6	5	4 (circled, boxed)	6	7
A	*III*	2 (circled)	4	3	3	8 (boxed)
	IV	5	6 (boxed)	2	2	1 (circled)

The solution of the game is

(i) The best strategy for *A* is II

(ii) The best strategy for *B* is III

(iii) The value of the game is $v = 4$.

9. In a game of matching coins with two players suppose A wins one unit of value when there are two heads, wins nothing when there are two tails and loses $\frac{1}{2}$ unit of value when there are one head and one tail. Determine the pay-off matrix, the best strategies for each player, and the value of the game to A.

[R. U. 87 H, Kuruk 77, Meerut 75, Kanpur 72, Delhi 64, 76, Rohilkhand 86, Andhra B. E. (Elect.) 83, V. B. U. Part III Hons. 97, 99 (Re-exam.)]

Solution :

The pay off matrix for *A* is

		B	
		H I (q_1) H	T *II* (q_2)
	I (p_1) H	+1 HH	$-\frac{1}{2}$ HT
A	*II* (p_2) T	$-\frac{1}{2}$ TH	0 TT

Clearly, the game does not have saddle point.

Here $\qquad a_{11} = 1, a_{12} = -\frac{1}{2}, a_{21} = -\frac{1}{2}$ and $a_{22} = 0$

$$p_1 = \frac{a_{22} - a_{21}}{(a_{11} + a_{22}) - (a_{12} + a_{21})} = \frac{1}{4}$$

$$p_2 = 1 - p_1 = 1 - \frac{1}{4} = \frac{3}{4} \text{ as } p_1 + p_2 = 1$$

$$q_1 = \frac{a_{22} - a_{12}}{(a_{11} + a_{22}) - (a_{12} + a_{21})} = \frac{1}{4}$$

$$q_2 = 1 - \frac{1}{4} = \frac{3}{4} \text{ as } q_1 + q_2 = 1$$

v = value of the game

$$= \frac{a_{11}\, a_{22} - a_{12}\, a_{21}}{(a_{11} + a_{22}) - (a_{12} + a_{21})} = -\frac{1}{8}$$

Thus, the optimal strategies are each player should show head $\frac{1}{4}$ of the time and tails $\frac{3}{4}$ of the time. The value of the game to A is $v = -\frac{1}{8}$.

It means A will loose on average $\frac{1}{8}$ unit each time the game is played.

Hence the game is unfair to A.

10. A and B play game in which each has three coins, $5p$, $10p$ and $20p$. Each player selects a coin without the knowledge of the other choice.

If the sum of the coin is an odd amount, A wins B's coin; if the sum is even, B wins A's coin.

Find the best strategy for each player and the value of the game.

[Meerut 76, 87, kanpur 82, Delhi (O R) 77, 87, V. B. U. Part III Hons. 96]

Solution :

For this problem, A's pay off matrix is as follows

		B		
		$5p$	$10p$	$20p$
	$5p$	−5	10	20
A	$10p$	5	−10	−10
	$20p$	5	−20	−20

It can be verified that the game has no saddle point. IIIrd row is dominated by IInd. So delete 3rd row. The reduced matrix is

	B		
	$5p$	$10p$	$20p$
$5p$	–5	10	20
$10p$	5	–10	–10

In this new matrix IIIrd column is dominated by II. Delete 3rd column. The reduced 2×2 matrix is

		B	
		$5p\ (q_1)$	$10p\ (q_2)$
	$5p\ (p_1)$	– 5	10
A	$10p\ (p_2)$	5	– 10

Here $a_{11} = -5,\ a_{12} = 10$

$a_{21} = 5,\ a_{22} = -10$

$$p_1 = \frac{a_{22} - a_{21}}{(a_{11} + a_{22}) - (a_{12} + a_{21})} = \frac{1}{2}$$

$$p_2 = 1 - \frac{1}{2} = \frac{1}{2} \text{ as } p_1 + p_2 = 1$$

$$q_1 = \frac{a_{22} - a_{12}}{(a_{11} + a_{22}) - (a_{12} + a_{21})} = \frac{2}{3}$$

$$q_2 = 1 - \frac{2}{3} = \frac{1}{3} \cdot \text{ as } q_1 + q_2 = 1$$

$$v = \frac{a_{11}\, a_{22} - a_{12}\, a_{21}}{(a_{11} + a_{22}) - (a_{12} + a_{21})} = 0$$

Thus, the optimal startegies are that none of the player should select the coin of $20p$; A should select coin of $5p$ and $10p$ equally i. e. for $\frac{1}{2}$ time while B should select coin of $5p$ for $\frac{2}{3}$ of time and coin of $10p$ for $\frac{1}{3}$ of the time.

The value of the game for A is $v = 0$ and B, $v = 0$.

If means each time the game is played neither A will gain nor B will loose any amount.

EXERCISE

1. Solve the following games

(i)

		B I	II	III	IV
	I	–5	3	1	20
	II	5	5	4	6
A	III	–4	–2	0	–5

Ans. A — II B — III $v = 4$.

[Kuruk (Start) 78, Roorkee (Applied Math) 73, 76]

(ii)

		B I	II	III
	I	–2	15	–2
A	II	–5	–6	–4
	III	–5	20	–8

Ans. A—I B—I or III $v = -2$

[Kuruk 79, Rohilkhand 81]

2. Solve the following games by the principle of Dominance

(i)

		B I	II	III
	I	1	8	3
A	II	6	4	5
	III	0	1	2

Ans. $\left(\frac{1}{4}, 0, \frac{3}{4}\right)$

for both players $v = \frac{7}{8}$

(Shivajee (Math) 76]

(ii)

		B I	II	III
	I	6	4	5
A	II	4	5	4
	III	6	3	1

Ans. $\left(\frac{1}{2}, \frac{1}{2}, 0\right)$ for A, $\left(0, \frac{1}{2}, \frac{1}{2}\right)$ for B, $v = \frac{9}{2}$

[Gujrat M. Sc. (Stat) 78]

7

Duality Theory

7.1. INTRODUCTION

Associated with every L. P. P. is always a corresponding L. P. P. called the Dual Problem, of the given L. P. P. The original L. P. P. is called the Primal Problem while the other is callled its Dual. It is important to note that, in general, either problem can be considered the primal with the remaining problem the dual. The optimal solution of either problem reveals in formation concerning the optimal solution of the other. If the optimal solution to one is known, then the optimal solution of the other is readly available.

7.2. DEFINITION OF SYMMETRIC DUAL PROBLEM

[Delhi (L. P.) 68, 69; agra 89; Meerut 73, 69, (Stat) 83, (L. P.) 68, 73 (S), Calicut 78. 76]

Let us consider a L. P. P. in the following form which will be called primal problem.

$$\left.\begin{aligned}
\text{Max.}\quad & Z_p = c_1x_1 + c_2x_2 + \dots\dots\dots + c_nx_n \\
\text{Subject to}\quad & \\
& a_{11}x_1 + a_{12}x_2 + \dots\dots\dots + a_{1n}x_n \le b_1 \\
& a_{21}x_1 + a_{22}x_2 + \dots\dots\dots + a_{2n}x_n \le b_2 \\
& \dots\dots\dots\dots\dots\dots\dots\dots\dots\dots\dots\dots \\
& \dots\dots\dots\dots\dots\dots\dots\dots\dots\dots\dots\dots \\
& a_{m1}x_1 + a_{m2}x_2 + \dots\dots\dots + a_{mn}x_n \le b_m \\
\text{and}\quad & x_1, x_2, \dots\dots\dots\dots\dots x_n \ge 0
\end{aligned}\right\} \quad \dots \text{(i)}$$

where the sign of all parameters (a, b, c's) are arbitrary.

Corresponding to this problem, the dual problem can be obtained by

(i) trnansposing the co-effcient matrix,

(ii) interchanging the role of constant terms and the co-effcient of objective function,

(iii) reverting the inequalities,

and (iv) minimizing the objective function instead of maximizing it.

The Dual Problem is as below

Min. $Z_d = b_1 w_1 + b_2 w_2 + \dots\dots\dots + b_m w_m$

subject to

$$\left.\begin{array}{l} a_{11}w_1 + a_{21}w_2 + \dots\dots\dots + am_1\, w_m \geq c_1 \\ a_{12}w_1 + a_{22}w_2 + \dots\dots\dots + a_{m2}\, w_m \geq c_2 \\ \dots\dots\dots\dots\dots\dots\dots\dots\dots\dots\dots\dots \\ \dots\dots\dots\dots\dots\dots\dots\dots\dots\dots\dots\dots \end{array}\right\} \quad \text{...(ii)}$$

$$a_{1n}\, w_1 + a_{2n}\, w_2 + \dots\dots\dots + a_{mn} w_n \geq c_n$$

and $\quad w_1, w_2, \dots\dots\dots w_m \geq 0$

(ii) is dual of (i), $w_1, w_2, \dots\dots\dots, w_m$ are called Dual variables.

Primal (i) and Dual (ii) can be remembered by using the following tabular from.

$$\begin{matrix} & (x_1 x_2 \dots\dots\dots x_n) & & \text{min.} \\ \begin{bmatrix} w_1 \\ w_2 \\ : \\ : \\ : \\ w_m \end{bmatrix} & \begin{bmatrix} a_{11} & a_{12} \dots\dots & a_{1n} \\ a_{21} & a_{22} \dots\dots & a_{2n} \\ : & : & : \\ : & : & : \\ : & : & : \\ a_{m1} & : & a_{mn} \end{bmatrix} & \leq & \begin{bmatrix} b_1 \\ b_2 \\ : \\ : \\ : \\ b_m \end{bmatrix} \end{matrix}$$

$$\geq \text{maxi } (c_1, c_2 \dots\dots\dots c_n)$$

Read horizontally for primal (i) and vertically for dual (ii)

Note–Using matrix notation the Primal and Dual can be put as following Primal Problem

Max. $\quad Z_p = C_x$

s. t. $\quad Ax \leq b$

and $\quad x \geq 0.$

Dual problem

Min. $\quad Z_d = b'w$

s. t. $\quad A'w \leq c'$

$\quad w \geq 0$

where $w = [w_1, w_2, \dots\dots\dots, w_m]$

A', b', c' are transposes of A, b & c respectively.

3. Un-symmetrical Dual Problem—

In matrix notation the Primal Dual can be written as max. $Z_p = C_x$

max. $Z_p = C_x$

s. t. $a_x = b$

and $x \geq 0$

Dual Problem

min. $Z_d = b'w$

subject to $A'w \geq c'$

$w \geq 0$

Note-Dual variables $w = [w_1, \ldots\ldots\ldots w_m]$ are unrestricted in sign.

4. Standard form of the Primal

(i) All the constraints involve the sign $\leq$, if it is a problem of maximization.

or (ii) All the constraints involve the sign $\geq$ if it is problem of minimization.

5. Dual of the dual of a given Primal problem is the primal.

[Raj (L. P.) 85, agra 89; Delhi (L. P.) 68, 69; Meerut (L. P.) 68, 73, 81, 84; Keral M. Sc. (Stat) 83; Kuruk 79, R.U. Part III Hons. 91; V. B. U. Part III Hons. 99]

Proof— Let as consider the L. P. P. Primal.

$$\left.\begin{array}{ll} \text{max.} & Z_p = c_1x_1 + c_2x_2 + \ldots\ldots\ldots + c_nx_n \\ \text{subject to} & a_{11}x_1 + a_{12}x_2 + \ldots\ldots\ldots + a_{1n}x_n \leq b_1 \\ & a_{21}x_1 + a_{22}x_2 + \ldots\ldots\ldots + a_{2n}x_n \leq b_2 \\ & a_{31}x_1 + a_{32}x_2 + \ldots\ldots\ldots + a_{3n}x_n \leq b_3 \\ & \ldots\ldots\ldots\ldots\ldots\ldots\ldots\ldots\ldots \\ & \ldots\ldots\ldots\ldots\ldots\ldots\ldots\ldots\ldots \\ & a_{m1}x_1 + a_{m2}x_2 + \ldots\ldots\ldots + a_{mn}x_n \leq b_m \\ \text{and} & x_1, x_2, \ldots\ldots\ldots, x_n \geq 0 \end{array}\right\} \quad \ldots\text{(i)}$$

Dual—The dual of (i) is as follows

$$\left.\begin{array}{ll} \text{Min.} & Z_d = b_1w_1 + b_2w_2 + \ldots\ldots\ldots + b_mw_n \\ \text{subject to} & a_{11}w_1 + a_{21}w_2 + \ldots\ldots\ldots + a_{m1}w_m \geq c_1 \\ & a_{12}w_1 + a_{22}w_2 + \ldots\ldots\ldots + a_{m2}wm \geq c_2 \\ & \ldots\ldots\ldots\ldots\ldots\ldots\ldots\ldots\ldots \\ & \ldots\ldots\ldots\ldots\ldots\ldots\ldots\ldots\ldots \\ & a_{1n}w_1 + a_{2n}w_2 + \ldots\ldots\ldots + -a_{mn}w_m \geq c_n \\ \text{and} & w_1, w_2, \ldots\ldots\ldots, w_n \geq 0 \end{array}\right\} \quad \ldots\text{(ii)}$$

Now, we put (ii) is standard form of primal by multiplying through by – 1.

$$\left.\begin{array}{ll} \text{Max.} & (-Z_d) = -b_1w_1 - b_2w_2 - \ldots\ldots\ldots - b_mw_m \\ \text{subject to} & -a_{11}w_1 - a_{21}w_2 - \ldots\ldots\ldots - a_{m1}w_m \leq -c_1 \\ & -a_{11}w_1 - a_{22}w_2 - \ldots\ldots\ldots - a_{m2}w_m \leq -c_2 \\ & \ldots\ldots\ldots\ldots\ldots\ldots\ldots\ldots\ldots \\ & \ldots\ldots\ldots\ldots\ldots\ldots\ldots\ldots\ldots \\ & -a_{1n}w_1 - a_{2n}w_2 \ldots\ldots\ldots\ldots - a_{mn}w_n \leq -c_n \\ \text{and} & w_1, w_2, \ldots\ldots\ldots, w_n \geq 0 \end{array}\right\} \quad \ldots\text{(iii)}$$

Now, let us consider (iii) as primal.
Then its dual is given by

$$\left.\begin{array}{ll} \text{Min.} & Z_v = -c_1v_1 - c_2v_2 - \ldots\ldots - c_nv_n \\ \text{Subject to} & -a_{11}v_1 - a_{12}v_2 - \ldots\ldots\ldots - a_{1m}v_n \geq -b_1 \\ & -a_{21}v_1 - a_{22}v_3 - \ldots\ldots\ldots a_{2m}v_n \geq -b_2 \\ & \ldots\ldots\ldots\ldots\ldots\ldots\ldots\ldots\ldots\ldots \\ & \ldots\ldots\ldots\ldots\ldots\ldots\ldots\ldots\ldots\ldots \\ & -a_{m1}v_1 - a_{m2}v_2 - \ldots\ldots\ldots -a_{2m}v_n \geq -b_m \\ \text{and} & v_1, v_2, \ldots\ldots\ldots v_n \geq 0 \end{array}\right\} \quad \ldots \text{(iv)}$$

Now, multiplying (iv) through by – 1, we get

$$\begin{array}{ll} \text{Max.} & Z_v = \min. (-Z_v) \\ & = c_1v_1 + c_2v_2 + \ldots\ldots\ldots + c_nv_n \\ \text{subject to} & a_{11}v_1 + a_{12}v_2 + \ldots\ldots\ldots + a_{1n}v_n \leq b_1 \\ & a_{21}v_1 + a_{22}v_2 + \ldots\ldots\ldots + a_{2n}v_n \leq b_2 \\ & \ldots\ldots\ldots\ldots\ldots\ldots\ldots\ldots\ldots\ldots \\ & \ldots\ldots\ldots\ldots\ldots\ldots\ldots\ldots\ldots\ldots \\ & a_{m1}v_1 + a_{m2}v_2 + \ldots\ldots\ldots + a_{mn}v_n \leq b_n \\ \text{and} & v_1, v_2, \ldots\ldots\ldots, v_n \geq 0 \end{array} \quad \ldots \text{(v)}$$

which is the system identical to the primal (i).

This completes the proof of the theorem.

ILLUSTRATIVE EXAMPLES

1. Obtain the dual of the following L. P. Problem.

$$\begin{array}{ll} \text{Mix.} & Z = 2x_2 + 5x_3 \\ \text{subject to} & x_1 + x_2 \geq 2 \\ & 2x_1 + x_2 + 6x_3 \leq 6 \\ & x_1 - x_2 + 3x_3 = 4 \\ & x_1, x_2, x_3 \geq 0 \end{array}$$

[Meerut (B. Sc.) Hons. 82]

Solution—

First we write the given problem into standard form. For this objective function is written as

$$\begin{array}{ll} \text{Max.} & Z_p = \min. (-Z) = -2x_2 - 5x_3 \\ & x_1 + x_2 \geq 0 \text{ is written as } -x_1 - x_2 + 0.x_3 \leq -2 \\ & x_1 - x_2 + 3x_3 = 4 \text{ can be expressed as} \\ & \left.\begin{array}{l} x_1 - x_2 + 3x_3 \leq 4 \\ x_1 - x_2 + 3x_3 \geq 4 \end{array}\right\} \\ \text{or,} & x_1 - x_2 + 3x_3 \leq 4 \\ \text{and} & -x_1 + x_2 - 3x_3 \leq -4 \end{array}$$

Then, the original problem now becomes of the standard primal form.

$$\begin{array}{ll} \text{Max.} & Z_p = 0.x_1 - 2x_2 - 5x_3 \\ & -x_1 - x_2 + 0.x_3 \leq 2 \end{array}$$

$$\left.\begin{array}{l}2x_1 + x_2 + 6x_3 \le 6 \\ x_1 - x_2 + 3x_3 \le 4 \\ -x_1 + x_2 - 3x_3 \le -4\end{array}\right\} \quad \ldots\ldots \text{(ii)}$$

$x_1, x_2, x_3 \ge 0$

Now, we apply the rules mentained in section 2 for getting the dual. The required Dual is given by

$\text{Min.} z_d = -2w_1 + 6w_2 + 4w_3 - 4w_4$

Subject to $-1w_1 + 2w_2 + 1w_3 - 1w_4 \ge 0$

$-1w_1 + 1w_2 - 1w_3 + 1w_4 \ge -2$ ...... (ii)

$0w_1 + 6w_2 + 3w_3 - 3w_4 \ge -5$

$w_1, w_2, w_3, w_4 \ge 0$

Put $w_3' = w_3 - w_4$

Then $\text{Min. } Z_d = 2w_1 + 6w_2 + 4w_3'$

s. t. $-1w_1 + 2w_2 + w_3' \ge 0$

$-1w_1 + 1w_2 - w_3' \ge -2$

$0w_1 + 6w_2 + 3w_3' \ge -5$

$w_1, w_2 \ge 0$ and w_3' unrestricted in sign.

2. Obtain the dual of the following L. P. problem

Max. $Z = 2x_1 + 3x_2 + x_3$

Subject to the constraints

$4x_1 + 3x_2 + x_3 = 6$

$x_1 + 2x_2 + 5x_3 = 4$

$x_1, x_2, x_3 \ge 0$

[Meerut B. Sc. (Math) 69, Punjab M. Sc. (Math) 75, Raj. M. Sc. 73,
T. D. C. II year 77 (N. C.) R. U. Condensed Course 89]

Solution —

The equation $4x_1 + 3x_2 + x_3 = 6$ is equivalent to

$4x_1 + 3x_2 + x_3 \le 6$ ... (i)

and $4x_1 + 3x_2 + x_3 \ge 6$

i. e. $-4x_1 - 3x_2 - x_3 \le -6$ ... (ii)

In the same way

$x_1 + 2x_2 + 5x_3 = 4$ is equivalent to

$x_1 + 2x_2 + 5x_3 \le 4$ ... (iii)

and $x_1 + 2x_2 + 5x_3 \ge 4$

i. e. $-x_1 - 2x_2 - 5x_3 \le -4$ ... (iv)

Now, we write the given problem in the standard primal form with the aid of (i), (ii), (iii) and (iv) as

$\text{Max. } Z_p = 2x_1 + 3x_2 + x_3$

Subject to $4x_1 + 3x_2 + x_3 \le 6$

$-4x_1 - 3x_2 - x_3 \le -6$

$x_1 + 2x_2 + 5x_3 \le 4$

$-x_1 - 2x_2 + 5x_3 \le -4$

$x_1, x_2, x_3 \ge 0$

Following the rules its dual is obtained as

Min. $Z_d = 6w_1 - 6w_2 + 4w_3 - 4w_4$

Subject to $4w_1 - 4w_2 + w_3 - w_4 \geq 2$

$3w_1 - 3w_2 + 2w_3 - 2w_4 \geq 3$ ... (i)

$w_1 - w_2 + 5w_3 - 5w_4 \geq 1$

$w_1 , w_2 , w_3 , w_4 \geq 0$

(i) Can be written as

Min. $Z_d = 6 (w_1 - w_2) + 4 (w_3 - w_4)$

Subject to $4 (w_1 - w_2) + (w_3 - w_4) \geq 2$

$3 (w_1 - w_2) + 2 (w_3 - w_4) \geq 3$ ...(ii)

$(w_1 - w_2) + 5 (w_3 - w_4) \geq 1$

$w_1 , w_2 , w_3 , w_4 , \geq 0$

Put $w_1 - w_2 = y_1$ and $w_3 - w_4 = y_2$

Then (ii) can be written as

Min. $Z_d = 6y_1 + 4y_2$

Subject to $4y_1 + y_2 \geq 2$

$3y_1 + 2y_2 \geq 3$

$y_1 + 5y_2 \geq 1$

y_1 , y_2 unrestricted in sign.

3. Find the dual of the following L. P. P.

Min. $Z = x_1 + x_2 + x_3$

Subject to $x_1 - 3x_2 + 4x_3 = 5$

$x_1 - 2x_2 \leq 3$

$2x_2 - x_3 \geq 4$

$x_1 , x_2 \geq 0$ and x_3 is unrestricted in sign.

[Meerut L. P. 87; Delhi M. A. (Bus. Eco) 78, B. Sc. (Math) 75]

Solution —

Put $x_3 = x_3' - x_3''$

Where $x_3' , x_3'' \geq 0$

Then L. P. P. becomes

Min. $Z = x_1 + x_2 + x_3' - x_3''$ ...(i)

Subject to $x_1 - 3x_2 + 4x_3' - 4x_3'' = 5$ ...(ii)

$x_1 - 2x_2 \leq 3$ ...(iii)

$2x_2 - x_3' + x_3'' \geq 4$ ...(iv)

$x_1 , x_2, x_3' , x_3'' \geq 0$

(i) Can be written as

Max. $Z_p = \min. (-Z) = -x_1 - x_2 - x_3' + x_3''$

(ii) is equivalent to

$x_1 - 3x_2 + 4x_3' - 4x_3'' \leq 5$

and $x_1 - 3x_2 + 4x_3' - 4x_3'' \geq 5$

or, $-x_1 + 3x_2 - 4x_3' + 4x_3'' \leq 5$

(iii) is $-2x_2 + x_3' - x_3'' \leq -4$

The given L. P. P. is standard primal as

Max. $Z_p = -x_1 - x_2 - x_3' + x_3''$

Subject to $x_1 - 3x_2 - 4x_3' + 4x_3'' \leq 5$

$-x_1 + 3x_2 - 4x_3' + 4x_3'' \leq -5$

$$x_1 - 2x_2 \le 3 \quad \ldots \text{(i)}$$
$$-2x_2 + x_3' - x_3'' \le -4$$
$$x_1, x_2, x_3', x_3'' \ge 0$$

Dual of (i) is

$$x_1, x_2, x_3', x_3'' \ge 0$$

Min. $Z_d = 5w_1 - 5w_2 + 3w_3 - 4w_4$

Subject to
$$1w_1 - 1w_2 + 1w_3 + 0.w_4 \ge -1$$
$$-3w_1 + 3w_2 - 2w_3 - 2w_4 \ge -1$$
$$4w_1 - 4w_2 + 0.w_3 - 1w_4 \ge -1$$
$$-4w_1 - 4w_2 + 0w_3 - 1w_4 \ge 1$$
$$w_1, w_2, w_3, w_4, \ge 0$$

Put $w_1 - w_2 = w_1'$. Then

Min. $Z_d = 5w_1' + 3w_3 - 4w_4$

Subject to
$$w_1' + w_2 \ge -1$$
$$-3w_1' - 2w_3 + 2w_4 \ge -1 \quad \ldots \text{(ii)}$$
$$4w_1' + w_4 \ge -1$$
$$4w_1' - w_4 \ge 1$$
$w_3, w_4 \ge 0$ and w_1' is unrestricted in sign.

(ii) Can also be written as

Max. $Z_d = -5w_1' - 3w_3 + 4w_4$

Subject
$$-w_1' - w_3 \le 1$$
$$3w_1' + 2w_3 + 2w_4 \le 1 \quad \ldots \text{(iii)}$$
$$-4w_1' - w_4 \le 1 \quad \ldots \text{(A)}$$
$$4w_1' + w_4 \le -1 \quad \ldots \text{(B)}$$
$w_3 \ge 0, w_4 \ge 0, w_1'$ unrestricted in sign.

Further (B) can be written as
$$-4w_1' - w_4 = 1 \quad \ldots \text{(C)}$$

The equivalent form of (A) and (C) is
$$4w_1' - w_4 = 1$$
i. e.
$$4w_1' + w_4 = -1$$

So (iii) be written as

Max. $Z_d = -5w_1' - 3w_3 + 4w_4$

Subject to
$$-w_1' - w_3 \le 1$$
$$3w_1' + 3w_3 + 2w_4 \le 1$$
$$4w_1' + w_4 = -1$$
$w_3 \ge 0, w_4 \ge 0, w_1'$ unrestricted in sign.

4. Obtain in the Dual of the following L. P. P.

Min. $Z = 2x_1 + 3x_2 + 4x_3$

Subject to
$$2x_1 + 3x_2 + 5x_3 \ge 2$$
$$2x_1 + x_2 + 7x_3 = 3$$
$$x_1 + 4x_2 + 6x_3 \le 5$$
$x_1, x_2 \ge 0$ and x_3 is unrestricted in sign.

[Gujrat (Stat) 78, Delhi (M. Com.) 78]

Solution —

Put $x_3 = x_3' - x_3''$ where $x_3', x_3'' \ge 0$

The standard primal form of given L. P. P. is

Max. $Z_p = \text{min.} (-Z) = -2x_1 - 3x_2 - 4(x_3' - x_3'')$

Subject to $-2x_1 + 3x_2 - 5(x_3' - x_3'') \geq -2$

$3x_1 + x_2 + 7(x_3' - x_3'') \leq 3$

$3x_1 - x_2 - 7(x_3' - x_3'') \leq -3$

$x_1 + 4x_2 + 6(x_3' - x_3'') \leq 5$

$x_1, x_2, x_3', x_3'' \geq 0$

Applying the rules its dual is

Min. $Z_d = -2v_1 + 3(v_2' - v_2'') + 5v_3$

Subject to $-2v_1 + 3(v_2' - v_2'') + v_3 \geq -2$

$-3v_1 + (v_2' - v_2'') + v_3 \geq -2$

$-5v_1 + 7(v_2' - v_2'') + 6v_3 \geq -4$

$5v_1 - 7(v_2' - v_2'') - 6v_3 \geq 4$

$v_1, v_2', v_2'', v_3 \geq 0$

Or, Min. $Z_d = -2v_1 + 3v_2 + 5v_3$

s. t. $-2v_1 + 3v_2 + v_3 \geq -2$

$-3v_1 + v_2 + 4v_3 \geq -3$

$-5v_1 + 7v_2 + 6v_3 = 4$, $v_1, v_3 \geq 0$, v_2 is unrestricted is sign.

EXERCISE

1. Write the dual of the following L. P. P.

Min. $Z = 3x_1 + 2x_2 + 4x_3$

s. t. $3x_1 + 5x_2 + 4x_3 \geq 4$

$6x_1 + x_2 + 3x_3 \geq 4$

$7x_1 - 2x_2 - x_3 \geq 10$

$x_1 - 2x_2 + 5x_3 \geq 3$

$4x_1 + 7x_2 + 2x_3 \geq 2$

$x_1, x_2, x_3 \geq 0$

[Dibrugrah (Stat) 74, Meerut B. Sc. (Math) 70]

Ans. Min. $Z_d = -7w_1 - 4w_2 + 10w_3 - 3w_4 - 2w_5$

s. t. $-3w_1 - 6w_2 + 7w_3 - w_4 - 4w_5 \geq -3$

$-5w_1 - w_2 - 2w_3 + 2w_4 - 7w_5 \geq 2$

$-4w_1 - 3w_2 - w_3 - 5w_4 + 2w_5 \geq -4$

$w_1, w_2, w_3, w_4, w_5 \geq 0$

2. Give the Dual of the following L. P.P.

Max. $Z = 3x_1 - 2x_2$

$x_1 \leq 4$

$x_2 \leq 6$

$x_1 + x_2 \leq 5$

$-x_2 \leq -1$

$x_1, x_2 \geq 0$

[Kuruk 76, Meerut L. P. 73 (S), 76, 77, 89]

Ans. Min. $Z = 4w_1 + 6w_2 + 5w_3 - 1w_4$

$$1w_1 + 0w_2 + 1w_3 + 0w_4 \geq 3$$
$$0w_1 + 1w_2 + 1w_3 - 1w_4 \geq -2$$
$$w_1, w_2, w_3, w_4 \geq 0$$

3. Convent the following L. P. P. to its dual problem

Min. $Z = 2x_1 + 2x_2 + 4x_3$

s.t

$$2x_1 + 3x_2 + 5x_3 \geq 2$$
$$3x_1 + x_2 + 7x_3 \leq 3$$
$$x_1 + 4x_2 + 6x_3 \leq 5$$
$$x_1 , x_2 , x_3 , \geq 0$$

[Meerut 75 (S)]

Ans. Min. $Z_d = -2w_1 + 3w_2 + 5w_3$

s. t.

$$-2w_1 + 3w_2 + w_3 \geq -2$$
$$-3w_1 + 1w_2 + 4w_3 \geq -2$$
$$-5w_1 + 7w_2 + 6w_3 \geq -4$$
$$w_1 , w_2 , w_3 \geq 0$$

4. Write the dual of the following L. P. P.

Min. $Z = 3x_1 + x_2$

s. t.

$$2x_1 + 3x_2 \geq 2$$
$$x_1 + x_2 \geq 1$$
$$x_1 , x_2 \geq 0$$

[Meerut 68, Delhi 68]

Ans. $Z_d = 2w_1 + w_2$

s. t.

$$2w_1 + w_2 \leq 3$$
$$3w_1 + w_2 \leq 1$$
$$w_1 , w_2 \geq 0$$

8

Degeneracy

1. **Definition**—A basic feasible solution of L.P.P. is said to be degenerate Basic feasible solution, if at least one of the basic variable is zero.

The phenomenon of getting a degenerate B.F.S. to L.P.P. is called **Degeneracy** in Linear programming.

2. **Cycling and Resolution of the degeneracy problem**

The presence of degeneracy occurs while applying the *Minimum ratio rule* $\left(\text{i.e. } \dfrac{X_B}{\text{Incoming vector}}\right)$ to determine the *outgoing vector*.

This possibility occurs whenever the *Minimum Ratio* is not unique. Then the problem is to choose the vector to be deleted from the basis. In the such cases, if we select one vector and delete it from the basis, then the next solution (obtained at this iteration) may be a B.F.S. This type of problem is called the *Problem of degeneracy*.

It can be seen that when the simplex method is applied to a degenerate B.F.S. to get a new B.F.S. the value of the objective function is not improved. In some cases due to the presence of degeneracy the same sequence of simplex table are repeated forever without ever reaching the optimal solution. This problem is called Cycling. When the successive iteration in a simplex method enter a loop which repeats the same sequence of iteration then this phenomenon is called *cycling in linear programming.*

"The procedure which prevent cycling within the simplex routine and the optimal solution is obtained in finite number of steps is called the Resolution of Degeneracy Problem.

3. Conditions for the appearance of degeneracy in a L.P.P.

[Raj L.P. 80, 81, 87: Meerut 72, 86, 87, 89: R.U. Part III Hons. 91]

(i) The degeneracy may appear in L.P.P. at the very first iteration when same component of vector b is zero.

(ii) If none of the components vector b is zero at any iteration and choice of out going vector at the same iteration is not unique, the next solution is bound to be degenerate.

4. Procedure for Computation

Let α_k be incoming vector in the simplex table and minimum ratio $\left(= \dfrac{X_B}{\text{Incoming vector}} \right)$ is not unique. In this case following procedure is adopted.

(i) Relabel the columns of the simplex table starting with the columns in the basis. Now, let the new columns be $\bar{Y}_1$, $\bar{Y}_2$ .... Let $\bar{Y}_t$ be the new number of entering vector Y_k (i. e. $Y_k = \bar{Y}_t$)

(ii) Let minimum ratio

$$\underset{t}{Min.} \left\{ \frac{Y_{B_i}}{Y_{tk}}, y_{tk} > 0 \right\}$$

appear at $i = i_1, \quad i_2, \quad i_3, \ldots\ldots i_r.$

Then, let $I_1 = \{i_1, i_2, i_3, \ldots\ldots\ldots\ldots i_r\}$

Compute $$\underset{t}{Min.} \left\{ \frac{\bar{y}_{ij}}{\bar{y}_{it}}, \bar{y}_{it} > 0 \right\} \forall\, i \in I_1$$

If the minimum is unique, then delete the corresponding vector from the basis. If the minimum is not unique, then proceed to the next step (iii).

(iii) Compute $$\underset{t}{Min.} \left\{ \frac{\bar{y}_{i2}}{\bar{y}_{it}}, \bar{y}_{it} > 0 \right\} \forall\, i \in I_2$$

where I_2 is the set of all those values of $I \in I_1$ for which there is a tie in (ii), clearly $I_2\, C I_1$.

If this minimum is unique, then delete the corresponding vector from the basis.

If this minimum is also not unique, then proceed to the next step (iv).

(iv) Compute $$\underset{t}{Min.} \left\{ \frac{\bar{y}_{i3}}{\bar{y}_{it}}, \bar{y}_{it} > 0 \right\} \forall\, i \in I_3$$

where I_3 is the set of these values of $i \in I_2$ for which there is a tie in (iii). Clearly $I_3 \subset I_2 \subset I_1$.

Continuing in this way we get a unique minimum value of i, i.e. the unique vector to be deleted from the basis.

For clear understanding of the above procedure see the worked out example.

Solved Example

1. Solve the following L.P.P.

$$\text{Max. } Z = 2x_1 + 3x_2 + 10x_3$$

$$\text{Subject to } x_1 + 2x_3 = 0$$

$$x_2 + x_3 = 1$$

$$x_1, x_2, x_3 \geq 0$$

[Meerut 70]

Solution—

The given problem can be put in following form—

$$\text{Max. } Z = 2x_1 + 3x_2 + 10x_3$$

$$\text{Subject to } x_1 + 0 \cdot x_2 + 2x_3 = 0$$

$$0x_1 + 1 \cdot x_2 + 1 \cdot x_3 = 1$$

Taking x_3 (non-basic variable) = 0, we get $x_1 = 0$, $x_2 = 1$

Thus, the starting B.F.S. is $x_1 = 0$, $x_2 = 1$, $x_3 = 0$. This B.F.S. is degenerate as the basic variable $x_1 = 0$. We construct the starting simplex table as follows.

Table 1

B	C_B	X_B	Y_1	Y_2	Y_3	$Min.Ratio = \frac{X_B}{Y_3}$
Y_1	2	0	1	0	[2]	$\frac{0}{2} \rightarrow$
Y_2	3	1	0	1	1	
$Z = C_B X_B$		c_j	2	3	10	
$= 0 \times 2 + 3 \times 1 = 3$		Δ_j	0	0	3	
			↓ Outgoing vector		↑ Incoming vector	

$$\Delta_1 = C_1 - C_B Y_1$$
$$= 2 - (2, 3) \times (1, 0) = 0 \text{ etc.}$$

Since $\Delta_j = 0$ for all j, therefore $Z = 3$ is not optimal solution. We construct the second simplex table as follows

Table 2

B	C_B	X_B	Y_1	Y_2	Y_3	
Y_3	10	0	$\frac{1}{2}$	0	1	
Y_2	3	1	$-\frac{1}{2}$	1	0	
$Z = C_B X_B$		c_j	2	3	10	
$= 10 \times 0 + 3 \times 1 = 3$		Δ_j	$-\frac{3}{2}$	0	0	

$$\Delta_j = 2 - (10, 3) \times \left(\frac{1}{2}, -\frac{1}{2}\right) = -\frac{3}{2} \text{ etc.}$$

Here $\Delta_j \leq 0 \ \forall_j$, So $Z = 2$ is optimal solution

i. e. $Z_{\text{max.}} = 3$ for $x_1 = 0, \quad x_2 = 1, \quad x_3 = 0$

This optimal solution is degenerate as $x_1 = 0$.

2. Solve the following L.P.P.

$$\text{Max.} \quad Z = 2x_1 + 3x_2 + 10x_3$$
$$\text{Subject to} \quad x_1 - 2x_3 = 0$$
$$x_2 + x_3 = 1$$
$$x_1, x_2, x_3, \geq 0$$

[Meerut B.Sc. (H) 82; M.Sc. 70]

Solution:

Put the L.P.P. in the following form.

$$\text{Max.} \quad Z = 2x_1 + 3x_2 + 10x_3$$
$$\text{Subject to} \quad 1 \cdot x_1 + 0 \cdot x_2 - 2 \cdot x_3 = 0$$
$$0 \cdot x_1 + x_2 + x_3 = 1$$

Taking $x_3 = 0$, we get

$$x_1 = 0, \quad x_2 = 1$$

Thus, starting B.F.S. is $x_1 = 0, x_2 = 1, x_3 = 0$

Which is clearly degenerate as the basic variable $x_1 = 0$.

The starting simplex table is as follows

Table 1

B	C_B	X_B	Y_1	Y_2	Y_3	Min. Ratio $= \frac{X_B}{Y_3}$
Y_1	2	0	1	0	−2	negative
Y_2	3	1	0	1	$\boxed{1}$	$\frac{1}{1} \rightarrow$
$Z = C_B X_B$		c_j	2	3	10	
$= 2 \times 0 + 3 \times 1 = 3$		Δ_j	0	0	11	

↓ Outgoing vector (Y_1) ↑ Incoming vector (Y_3)

$$\Delta_1 = C_1 - C_B Y_1 = 2 - (2, 3) \times (1, 0) = 2 - 2 = 0$$
$$\Delta_2 = C_2 - C_B Y_2 = 3 - (2, 3) \times (0, 1) = 0$$
$$\Delta_3 = C_3 - C_B Y_3 = 10 - (2, 3)(-2, 1) = 10 - (-4 + 3)$$
$$= 10 - (-1) = 11$$

Since $\Delta_1 = \Delta_2 = 0$ and $\Delta_3 > 1$, therefore $Z = 3$ is not optimal solution, we form the second simplex table.

Table 2

B	C_B	X_B	Y_1	Y_2	Y_3	Min. Ratio
Y_1	2	2	1	2	0	
Y_3	10	1	0	1	1	
$Z = C_B X_B$		c_j	2	3	10	
$= 2 \times 2 + 10 \times 1 = 4 + 10 = 14$		Δ_j	0	−11	0	

Since $\Delta_j \leq 0, \quad \forall_j$, therefore $Z = 14$ is optimal solution.

$$Z_{\text{max.}} = 14 \text{ for } x_1 = 2, x_2 = 0, x_3 = 1$$

The optimal solution is non-degenerate.

3. Solve the following L.P. Problem,

$$\text{Max.} \quad Z = 5x_1 + 3x_2$$

Subject to
$$x_1 + x_2 \le 2$$
$$5x_1 + 2x_2 \le 10$$
$$3x_1 + 8x_2 \le 12$$
$$x_1, x_2 \ge 0$$

[Roorkee (Applied Math) 76, Meerut L.P. 74, 87]

Solution:

Convert the constraints into equations by introducing the slack variables x_3, x_4, x_5. Then the L.P.P. becomes

$$\text{Max.} \quad Z = 5x_1 + 3x_2$$

Subject to
$$x_1 + x_2 + x_3 = 2$$
$$5x_1 + 2x_2 + x_4 = 10$$
$$3x_1 + 8x_2 + x_5 = 12$$

Putting $x_1 = 0$, $x_2 = 0$, we get $x_3 = 2$, $x_4 = 10$, $x_5 = 12$ which is starting B.F.S.

We construct the simplex table as follows

Table 1

			$\bar{Y}_4$	$\bar{Y}_5$	$\bar{Y}_1$	$\bar{Y}_2$	$\bar{Y}_3$	Mini. ratio $= \frac{X_B}{Y_1}$
B	C_B	X_B	Y_1	Y_2	Y_3	Y_4	Y_5	
Y_3	0	2	1	1	1	0	0	$\frac{2}{1}$
Y_4	0	10	[5]	2	0	1	0	$\frac{10}{5} = 2 \rightarrow$
Y_5	0	12	3	8	0	0	1	$\frac{12}{3} = 4$
$Z = C_B X_B = 0 \times 2 + 0 \times 10 + 0 \times 12 = 0$		c_j	5	3	0	0	0	
		Δ_j	5	3	0	0	0	

↑ Incoming vector (under Y_1); ↓ Outgoing vector (under Y_4)

Calculate Δ by the formula

$$\Delta_j = c_j - C_B Y_j$$

$$\Delta_1 = c_1 - C_B Y_1 = 5 - (0, 0, 0) \times (1, 5, 3) = 5$$

$$\Delta_2 = c_2 - C_B Y_2 = 3, \Delta_3 = 0, \Delta_4 = 0, \Delta_5 = 0$$

$$\Delta_j \geq 0 \text{ for all } j. \text{ So solution is not optimal.}$$

Here, minimum ratio is not unique.

It appears in the rows of Y_3 and Y_4 . Thus the problem is the problem of degeneracy we see that the identity matrix.

$$I = \begin{matrix} Y_3 & Y_4 & Y_5 \\ \end{matrix} \begin{bmatrix} 1 & 0 & 0 \\ 0 & 1 & 0 \\ 0 & 0 & 1 \end{bmatrix}$$

begins with Y_3 . So renumber the columns of table 1 such that the columns associated to the identity matrix come first in the proper order, i.e.

Write $\overline{Y_1} = Y_3$, $\overline{Y_2} = Y_4$, $\overline{Y_3} = Y_5$

$\overline{Y_4} = Y_1$, $\overline{Y_5} = Y_2$

Incoming vector is Y_1 (= $\overline{Y_4}$)
To determine which one (Y_3 or Y_4) is to be deleted.
This can be done as below

$(\overline{Y_4})$	$(\overline{Y_1})$	
Y_1	Y_3	Take the columns of Y_1 and $\overline{Y_1}$ and the rows of Y_3 and Y_4 i.e. the rows in which minimum ratio 2 is found.
1	1	
5	0	

B	$(\overline{Y_4})$	$(\overline{Y_1})$	*Min. ratio* $\frac{\overline{Y_1}}{Y_1}$
Y_3	$\frac{1}{1}$	$\frac{1}{3}$	$\frac{1}{1} = 1$
Y_4	5	0	$\frac{0}{5} = 0 \rightarrow$

Thus Y_4 is outgoing vector as shown in table.

Since solution found in table 1 is not optimal, we construct the second table.

Table 2

B	C_B	X_B	Y_1	Y_2	Y_3	Y_4	Y_5	Mini.ratio $\frac{X_B}{Y_2}$
Y_3	0	0	0	$\boxed{\frac{3}{5}}$	1	$-\frac{1}{5}$	0	0 →
Y_4	5	2	1	$\frac{2}{5}$	0	$\frac{1}{5}$	0	5
Y_5	0	6	0	$\frac{34}{5}$	0	$-\frac{3}{5}$	1	$\frac{15}{17}$
$Z = C_B X_B$		c_j	5	3	0	0	0	
= 10		Δ_j	0	1	0	− 1	0	
				↑	↓			

Since some $\Delta_j \geq 0 \; \forall_j$, therefore this calls for fresh construction of table

B	C_B	X_B	Y_1	Y_2	Y_3	Y_4	Y_5	Mini. ratio
Y_2	3	0	0	1	$\frac{5}{3}$	$-\frac{1}{3}$	0	
Y_1	5	2	1	0	$-\frac{2}{3}$	$\frac{1}{3}$	0	
Y_5	0	6	0	0	$-\frac{34}{3}$	$\frac{5}{3}$	1	
$Z = C_B X_B$		c_j	5	3	0	0	0	
= 10		Δ_j	0	0	$-\frac{5}{3}$	$-\frac{2}{3}$	0	

Since $\Delta_j \leq 0 \; \forall_j$, therefore the solution is optimal.

$$Z_{max} = 10 \text{ for } x_1 = 2, x_2 = 0.$$

4. Solve the L.P.P. by simplex method.

$$\begin{array}{llll} \text{Max.} & Z & = & 2x_1 + x_2 \\ \text{Subject to} & & & 4x_1 + 3x_2 \leq 12 \\ & & & 4x_1 + x_2 \leq 8 \\ & & & 4x_1 - x_2 \leq 8 \\ \text{and} & & & x_1, x_2 \geq 0 \end{array}$$

[Meerut L. P. '89; Agra 89]

Solution :

Introduce the slack variable x_3 , x_4 and x_5.

Then Max. $\quad Z = 2x_1 + x_2$

Subject to

$$4x_1 + 3x_2 + x_3 = 12$$

$$4x_1 + x_2 + x_4 = 8$$

$$4x_1 - x_2 + x_5 = 8$$

$$x_1 , x_2, x_3, x_4 , x_5 \geq 0$$

Putting $x_1 = 0, x_2 = 0$, we have $x_3 = 12, x_4 = 8, x_5 = 8$ which is the starting B.F.S.

The first simplex table is

Table 1

B	C_B	X_B	$\bar{Y}_4$	$\bar{Y}_5$	$\bar{Y}_1$	$\bar{Y}_2$	$\bar{Y}_3$	*Mini. ratio*
			Y_1	Y_2	Y_3	Y_4	Y_5	$\frac{X_B}{Y_1}$
Y_3	0	12	4	3	1	0	0	3
Y_4	0	8	4	1	0	1	0	2
Y_5	0	8	$\boxed{4}$	-1	0	0	1	2 →
$Z = C_B X_B$		c_j	2	1	0	0	0	
$= 0$		Δ_j	2	1	0	0	0	

↑ Incoming vector $\qquad$ ↓ Outgoing vector

Minimum ratio occurs in the rows of Y_4 and Y_5 showing that it is not unique. So it is problem of degenerancy.

For outgoing vector we proceed as follows

$$\text{The identity matrix is } \begin{array}{c} Y_3 \; Y_4 \; Y_5 \\ \begin{bmatrix} 1 & 0 & 0 \\ 0 & 1 & 0 \\ 0 & 0 & 1 \end{bmatrix} \end{array}$$

Relabel the columns of table 1 such that the columns associated to the identity matrix come first in the proper order in the following way $Y_3 = \overline{Y_1}$, $Y_4 = \overline{Y_2}$, $Y_5 = \overline{Y_3}$, $Y_1 = \overline{Y_4}$, $Y_2 = \overline{Y_5}$

Take the columns of Y_1 and $\bar{Y}_1$ and the rows of Y_4 and Y_5 i. e. the rows in which the minimum ratio 2 occurs.

	$(\bar{Y}_4)$ Y_1	$(\bar{Y}_2)$ Y_3	Min. ratio $\frac{\bar{Y}_1}{Y_1}$
Y_4	4	0	$\frac{0}{4} = 0$
Y_5	4	0	$\frac{0}{4} = 0$

Again, minimum ratio is not unique and occurs in the rows of Y_4 and Y_5 . So we take Y_1 and the column. $\bar{Y}_2$ next to $\bar{Y}_1$

	$(\bar{Y}_4)$ Y_1	$(\bar{Y}_2)$ Y_4	Min. ratio $\frac{\bar{Y}_1}{Y_2}$
Y_4	4	1	$\frac{1}{4}$
Y_5	4	0	$\frac{0}{4} = 0 \rightarrow$
	↑		

Thus minimum ratio is in the row of Y_5 .

$\therefore$ Y_5 is outgoing vector as shown in table 1.

Since $\Delta_j \nless 0, \forall_j$ therefore the solution obtained in table 1 is not optimal.

Following the computational proceduere for simplex method, the following tables have been constructed.

Table 2

B	C_B	X_B	Y_1	Y_2	Y_3	Y_4	Y_5	Mini. ratio $\frac{X_B}{Y_2}$
Y_3	0	4	0	4	1	0	-1	1
Y_4	0	0	0	[2]	0	1	-1	$0 \rightarrow$
Y_1	2	2	1	$-\frac{1}{4}$	0	0	$\frac{1}{4}$	(negative)
$Z = C_B X_B$ $= 0 \times 4$ $+ 0 \times 0$		c_j	2	1	0	0	0	
$+ 2 \times 2 = 4$		Δ_j	0	$\frac{3}{2}$	0	0	$-\frac{1}{2}$	

Table 3

B	C_B	X_B	Y_1	Y_2	Y_3	Y_4	Y_5	Mini. ratio $\frac{X_B}{Y_5}$
Y_3	0	4	0	0	1	-2	$\boxed{1}$	$\frac{4}{1} \rightarrow$
Y_4	1	0	0	1	0	$\frac{1}{2}$	$-\frac{1}{2}$	negative
Y_1	2	2	1	0	0	$\frac{1}{8}$	$\frac{1}{8}$	16
$Z = C_B X_B$		c_j	2	1	0	0	0	
$= 4$		Δ_j	0	0	0	$-\frac{3}{4}$	$\frac{1}{4}$	
					↓		↑	

Table 4

B	C_B	X_B	Y_1	Y_2	Y_3	Y_4	Y_5	
Y_5	0	4	0	0	1	-2	1	
Y_4	1	0	0	1	$\frac{1}{2}$	$-\frac{1}{2}$	0	
Y_1	2	$\frac{3}{2}$	1	0	$-\frac{1}{8}$	$\frac{3}{8}$	0	
$Z = C_B X_B = 5$		c_j	2	1	0	0	0	
		Δ_j	0	0	$-\frac{1}{4}$	$-\frac{1}{2}$	0	

Since $\Delta_j \leq 0, \forall_j$, therefore $Z = 5$ is optimal.

$$Z_{\text{max.}} = 5 \text{ for } x_1 = \frac{3}{2}\ x_2 = 2$$

OBJECTIVE QUESTIONS

1. If a set contains only interior points then the set is called.
   (i) closed set (ii) open set
   (iii) neither open nor closed (iv) none of these
2. If there exists an $\in$-neighbourhood about **a** which contains only points of the set S, then the point **a** of S is called
   (i) an exterior point (ii) a boundary point
   (iii) an interior point (iv) none of these
3. A hyper plane divides the whole space R^{n} into there mutually disjoint sets given by
   $X_1 = \{x : cx > z\}$
   $X_2 = \{x : cx = z\}$
   $X_3 = \{x : cx < z\}$
   The sets x_1 and x_2 are called
   (i) closed half space (ii) open
   (iii) open half space (iv) none of these
4. The set of all convex combinations of finite number of points is called
   (i) the Convex function (ii) Convex hull,
   (iii) Convex poly hedron (iv) none of these
5. In R^n, if the two points x_1 and x_2, $x_1 \neq x_2$ is defined to be the sets of points
   $X = \{x : \lambda x_2 + (1 - \lambda)\, x_1$, for all real values of $\lambda\}$ is called
   (i) lines (ii) line segment
   (iii) not line (iv) none of these
6. The set of all feasible solutions of L.P.P. is a
   (i) Concave set (ii) Concave hull
   (iii) Convex hull (iv) Convex set
7. The intersection of two Convex sets is also
   (i) a Concave set (ii) a Convex set
   (iii) a open set (iv) none of these
8. A hyper plane is
   (i) Concave set (ii) Convex set
   (iii) Convex hull (iv) none of these
9. The area of a triangle is
   (i) a Convex polyhedron of its vertices,
   (ii) a Convex function
   (iii) a Convex hull
   (iv) none of these

10. A linear equation in two variables

$$x_1, x_2, \text{ that is, } a_1x_1 + a_2x_2 = b \text{ represents}$$

(i) a point in two dimension (ii) a line in two dimension
(iii) a plane in two dimension (iv) none of these

11. If a constraint has a sign $\geq$, then positive variables which are subtracted from the L.H.S. of the Constraints to Convert them into equalities are called
(i) The surplus variable (ii) The slack variable
(iii) The basic variable (iv) none of these

12. If $x_1 + 3x_2 \geq 7$ is converted into

$$x_1 + 3x_2 + x_3 = 7 \text{ then } x_3 \text{ is called}$$

(i) surplus variable (ii) slack variable
(iii) the basic variable (iv) the non-basic variable
(v) none of these

13. If $5x_1 + 6x_2 \geq 7$, is converted into
$5x_1 + 6x_2 - x_3 = 7$, then x_3 is called
(i) slack variable (ii) surplus variable
(iii) basic variable (iv) none of these

14. The non-zero variables in a basic feasible solutions are called
(i) basic variables (ii) non-basic variables
(iii) slack variables (iv) none of these

15. If the solution of

$$x_1 + 2x_2 + x_3 = 4$$
$$2x_1 + x_2 + 5x_3 = 5$$

is
(a) Basic $x_1 = 2, x_2 = 1$ (b) non-basic $x_3 = 0$,
then the solution is called
(i) degenerate basic feasible (ii) non-degenerate basic feasible
(iii) none of these

16. If in the solution of a L.P.P. by simplex method all of Δ_j are negative, the solution under test is
(i) unique optimal solution (ii) not unique optimal solution

17. If the net evaluation row $\Delta_j \leq 0$ for each j, then the solution under test is
(i) optimal (ii) not optimal

18. In R^n

    $x_1^2 + x_2^2 \text{ ....} + x_n^2 = 1$ represents.

    (i) Straight line (ii) Plane
    (iii) Circle (iv) Hypersphere

19. If $\vec{a} = \lambda_1 \overrightarrow{a_1} + \lambda_2 \overrightarrow{a_2} + \ldots\ldots + \lambda_n \overrightarrow{a_n}$ is a linear convex combination of $a_1, a_2, \ldots, a_n$, then

    (i) $\lambda_r \le 0, r = 1, 2, \ldots, n$
    (ii) $\lambda_r \ge 0, r = 1, 2, \ldots, n$
    (iii) $\lambda_r \ge 0$ and $\sum_{r=1}^{n} \lambda_r = 1, r = 1, 2, \ldots, n$
    (iv) $\sum_{r=1}^{n} \lambda_r = 0$

20. The convex hull of the set

    $$x = \{(x_1, x_2) \mid x_1 x_2 \le 1, x_1 \ge 0, x_2 \ge 0\} \text{ is}$$

    (i) whole first quadrant (ii) whole 2nd quadrant
    (iii) whole 3rd quadrant (iv) whole 4th quadrant

21. Let $\vec{Z}(\lambda) = \lambda \vec{a} + (1 - \lambda) \vec{b} \in S \subseteq R^n$, where $\vec{a}, \vec{b} \in S$. Then S is a convex set if

    (i) $0 < \lambda \le 1$ (ii) $0 \le \lambda < 1$
    (iii) $0 \le \lambda \le 1$ (iv) $1 \le \lambda \le 2$

22. Which one is not convex set?

    (i) Sphere (ii) Hyperplane
    (iii) The closed half space (iv) The set $x = \{(x_1, x_2) \mid x_1^2 + x_2^2 = 1\}$

23. Vogel Approximation method is used to obtain the solution of

    (i) Transportation problem (ii) Assignment problem
    (iii) Dual problem (iv) Degeneracy problem

24. In a Balanced Transportation problem

    (i) $\sum_{i=1}^{m} a_i = \sum_{j=1}^{n} b_j$ (ii) $\sum_{i=1}^{m} a_i \ne \sum_{j=1}^{n} b_j$
    (iii) $\sum_{i=1}^{m} a_i > \sum_{j=1}^{n} b_j$ (iv) $\sum_{i=1}^{m} a_i < \sum_{j=1}^{n} b_j$

    where a_i, b_j denote supply and demand respectively.

25. A feasible solution of m by n transportation problem is said to be Basic feasible solution if the total number of positive allocations is exactly equal to

    (i) $m - n$ (ii) $m + n - 1$
    (iii) $m + n - 2$ (iv) $m + n$

26. In an assignment problem matrix

(i) $m = n$ (ii) $m < n$

(iii) $m > n$ (iv) $m = 2n$

where m is number of rows and n is number of columns

27. If number of decision variables is 3 and number of constraints is 2, then the number of basic solutions is

(i) 4 (ii) 1

(iii) 2 (iv) 3

28. In the following game

		B	
A	I	10	15
	II	3	2

the value of game for A is

(i) 10 (ii) 15

(iii) 3 (iv) 2

29. For the following game

		B: I	II
A	I	7	10
	II	5	12
	III	2	6

the best strategy for A is

(i) I (ii) II

(iii) III (iv) none of these

30. In the following game

		B: I	II	III
A	I	6	8	6
	II	4	12	2

value of game for B is

(i) – 6 (ii) + 6

(iii) – 12 (iv) – 4

31. A saddle point is a position of such an element in the pay-off matrix is

(i) minimum in its row and minimum in its column

(ii) maximum in its row and maximum in its column

(iii) minimum in its rows and maximum in its column

(iv) maximum in its row and minimum in its column

32. If n be the number of decision variables and m the number of constraints and $m < n$, then the number of basic solution is

(i) ${}^nC_m$ (ii) ${}^mC_n$

(iii) ${}^mC_m$ (iv) ${}^nC_n$

33. A feasible solution of m by n transportation problem is said to be Non-degenerate basic feasible solution if total number of positive allocations is exactly equal to

(i) $m + n - 1$ and allocations are independent of position

(ii) $m + n - 1$ and allocations are dependent of position

(iii) $m + n + 1$ and allocations are independent of position

(iv) $n - m + 1$ and allocations are independent of position

34. For solution of a Transportation problem penalty are required in the

(i) Hungarian Method (ii) V.A.M. Method

(iii) North-West Corner Rule (iv) Lowest Cost Entry Method

35. Cost of transportation of one unit from ith source to the jth destination is denoted by

(i) C_{ii} (ii) C_{jj}

(iii) C_{ij} (iv) C_{ji}

36. The quantity transported from ith source to jth destination is denoted by

(i) x_{ii} (ii) x_{ij}

(iii) x_{jj} (iv) x_{ji}

ANSWERS

1. (ii)	2. (iii)	3. (iii)	4.(iii)
5. (i)	6. (ii)	7. (ii)	8. (ii)
9. (i)	10. (ii)	11. (i)	12.(ii)
13. (ii)	14. (i)	15. (ii)	16.(i)
17. (i)	18. (iv)	19. (iii)	20.(i)
21. (iii)	22. (iv)	23. (i)	24.(i)
25. (ii)	26. (i)	27. (iv)	28 (i)
29. (i)	30. (i)	31. (iii)	32.(i)
33. (i)	34. (ii)	35. (iii)	36.(ii)